Brazilian Bodies and Their Choreographies of Identification

New World Choreographies

Series Editors: **Rachel Fensham** and **Peter M. Boenisch**

Editorial Assistant: **Alexandra Heller-Nicholas**

New World Choreographies presents advanced yet accessible studies of a rich field of new choreographic work which is embedded in the global, transnational and intermedial context. It introduces artists, companies and scholars who contribute to the conceptual and technological rethinking of what constitutes movement, blurring old boundaries between dance, theatre and performance.

The series considers new aesthetics and new contexts of production and presentation, and discusses the multi-sensory, collaborative and transformative potential of these new world choreographies.

Gretchen Schiller & Sarah Rubidge (*editors*)
CHOREOGRAPHIC DWELLINGS

Prarthana Purkayastha
INDIAN MODERN DANCE, FEMINISM AND TRANSNATIONALISM

Royona Mitra
AKRAM KHAM

Forthcoming titles include:

Pil Hansen & Darcey Callison (*editors*)
DANCE DRAMATURGY

New World Choreographies
Series Standing Order ISBN 978–1–137–35986–5 (hardback)
(*outside North America only*)

You can receive future titles in this series as they are published by placing a standing order. Please contact your bookseller or, in case of difficulty, write to us at the address below with your name and address, the title of the series and the ISBN quoted above.

Customer Services Department, Macmillan Distribution Ltd, Houndmills, Basingstoke, Hampshire RG21 6XS, England

Brazilian Bodies and Their Choreographies of Identification

Swing Nation

Cristina F. Rosa

palgrave
macmillan

First published 2015 by
PALGRAVE MACMILLAN

Palgrave Macmillan in the UK is an imprint of Macmillan Publishers Limited, registered in England, company number 785998, of Houndmills, Basingstoke, Hampshire RG21 6XS.

Palgrave Macmillan in the US is a division of St Martin's Press LLC, 175 Fifth Avenue, New York, NY 10010.

Palgrave Macmillan is the global academic imprint of the above companies and has companies and representatives throughout the world.

Palgrave® and Macmillan® are registered trademarks in the United States, the United Kingdom, Europe and other countries.

ISBN 978–1–137–46226–8

This book is printed on paper suitable for recycling and made from fully managed and sustained forest sources. Logging, pulping and manufacturing processes are expected to conform to the environmental regulations of the country of origin.

A catalogue record for this book is available from the British Library.

Library of Congress Cataloging-in-Publication Data
Rosa, Cristina F.
Brazilian bodies and their choreographies of identification : swing nation / Cristina F. Rosa.
pages cm
ISBN 978–1–137–46226–8 (hardback)
1. Capoeira (Dance) 2. Capoeira (Dance)—Social aspects—Brazil. 3. Dance—Social aspects—Brazil. 4. Choreography. I. Title.
GV1796.C145R67 2015
793.3'1981—dc23 2015005432

Typeset by MPS Limited, Chennai, India.

Contents

List of Figures

Series Editors' Preface

Choreography in the global context of the twenty-first century involves performance practices that are often fluid, mediated, interdisciplinary, collaborative, and interactive. Choreographic projects and choreographic thinking circulate rapidly within the transnational flows of contemporary performance, prompting new aesthetics and stretching the disciplinary boundaries of established "dance studies." Crossing the borders of arts disciplines, histories, and cultures, these "new world choreographies" utilize dance techniques and methods to new critical ends in the body's interaction with the senses, the adoption of technology, and the response to history, as well as present-day conditions of political and social transformation, or in its constitution of spectator communities.

As a result, well-rehearsed approaches to understanding choreography through dance lineages, canonical structures, or as the product of individual artists give way to new modes of production and representation and an ever-extending notion of what constitutes dance in performance. Choreographic practice as well as research on choreography draws on new methods of improvisation, (auto-)biography, collective creation, and immersion in ways which challenge established (Western) notions of subjectivity, of the artist as creator, or which unsettle the "objective distance" between the critic and the work. The post-national, inter-medial and interdisciplinary contexts of digital and social media, festival circuits, rapidly changing political economies, and global politics call for further critical attention.

With an openness to these new worlds in which dance so adeptly maneuvers, this book series aims to provide critical and historicised perspectives on the artists, concepts, and cultures shaping this creative field of "new world choreographies." The series will provide a platform for fresh ways to understand and reflect upon what choreography means to its various audiences, and to the wider field of international dance and performance studies. In addition, it will also provide a forum for new scholars to expand upon their ideas and to map out new knowledge paradigms that introduce this diverse and exciting field of choreographic practice to dance, theatre, and performance studies.

Rachel Fensham, University of Melbourne
Peter M. Boenisch, University of Kent
Alexandra Heller-Nicholas, Series Administration

We gratefully acknowledge the support of this publication by the faculty of Arts at the University of Melbourne.
www.newworldchoreographies.com

Acknowledgments

I would like to acknowledge the University of California, Los Angeles for providing support during my doctoral research, which culminated in this monograph. In particular, I would like to thank the Department of World Arts and Cultures/Dance, the Latin American Institute, the Center for the Study of Women, and the Department of Spanish and Portuguese at UCLA for financially sponsoring this project through a number of grants and fellowships.

For their generosity and attention during my multiple research trips to Brazil, I would like to thank Candida Braz, the archivist at Grupo Corpo, and Rodrigo Pederneiras, its resident choreographer; Mestre Valmir Damasceno, founder of the Fundação Internacional de Capoeira Angola-Bahia (FICA-BA); Helena Katz, the Director of Centro de Estudos em Dança at PUC-SP; Mestre Cobra Mansa, founder of the Kilombo Tenondé Center for Capoeira Angola and Permaculture in Bonfim, Bahia; Soraia da Silva, the Director of Centro de Pesquisa em Dança (CDPDan) at the Universidade de Brasília (UnB), and (the late) Fred Abreu, founder and former Director of the Instituto Jair Moura de Capoeira in Salvador, BA. In addition, I extend my gratitude to the Centro de Estudos Afro-Asiáticos (CEAO) at the Universidade Federal da Bahia (UFBA) for their contributions to this project during my research time in Salvador, Bahia.

For their critical feedback as I tried to get a handle on the archival and ethnographic data collected during the early stages of this project, I would like to extend special thanks to Professors Andrew Apter, Rosângela Araújo, Paula Barreto, Sue-Ellen Case, Donald Cosentino, David Gere, Randal Johnson, Allen Roberts, Livio Sansone, Carlos Eugenio L. Soares, Janet O'Shea, and José Miguel Wisnik. Above all, I would like to thank Susan Leigh Foster, my mentor and PhD advisor. Her impeccable ethics, unwavering commitment, and thoughtfulness have inspired me and kept me on track all along. Many thanks also to my colleagues at UCLA, with a special *obrigado* to Lorena Alvorado, Feriyal Aslam, Rosemary Candelario, Wes Days, Arianne Hoffman, CedarBough T. Saeji, D. Sabela Grimes, Carla Melo, Jose Reynoso, Michael Sakamoto, Carolina San Juan, Angeline Shaka, Raphael Xavier, Giavanni Washington, and Sara Wolf. I am certain that without their serious input and playful camaraderie, I could not gone this far.

For their continuous support as I transformed this research project into a book manuscript, I would like to acknowledge the incredible librarians and staff who assisted me in the following institutions: Florida State University, Tallahassee; the International Research Center "Interweaving Performance Cultures" at FU, Berlin; the International Foundation of Capoeira Angola in Washington DC; the University of California, Riverside; and Reed College. I am especially grateful for the kindness of these individuals and organizations, granting me access to their special collections and offering me personal assistance when needed. I am also thankful to Rachel Fensham and Peter M Boenisch, the editors of this book series, for offering their expertise and generous support. Likewise, I am especially obligated to Lauren Davidson, Jose Reynoso, Davi F. Rosa, Ana Paula F. Rosa, and Jonathan Wolf for their editorial assistance in different chapters, as well as Jon Lloyd, for painstakingly copy-editing the entire manuscript.

This project has benefited from a wide variety of audio-visual material, some of which are reproduced here. In this regard, I would like to acknowledge: the Biblioteca Brasiliana Guita e José Mindlin, USP and Acervo da Fundação Biblioteca Nacional, Brazil for kindly providing digital reproductions of public domain images housed at their archives; the dance company Grupo Corpo for cordially allowing me to attend its open rehearsals and photograph its activities all the samba and capoeira master teachers who have welcomed me in their "homes" and have cordially permitted me to document their classes and public events. I would especially like to thank treinel Adijair "Dija" Damasceno (FICA-BA), who is featured on the cover of this book, and Ugo Edu, who took the photograph.

Finally, I would like to thank my parents, Maria Célia and Darlan, for their unconditional love and trust.

Portions of this book were previously published in the following essays and I thank the presses for permission to reproduce them here:

- "Performing Brazil: The Case of Grupo Corpo," in *Performing Brazil: Essays on Culture, Identity, and the Performing Arts*. Madison: University of Wisconsin Press, 2015, pp. 67–97.
- "Playing, Fighting, and Dancing: Unpacking the Significance of Ginga within the Practice of Capoeira Angola," *The Drama Review*, 56(3) (2012): 141–66.

Introduction: Choreographing Ideas

As it so happened, one day in the morning, as I was strolling through the grounds of my suburban home, an idea took hold in the trapeze I carry about in my brain. Once hanging there, it began to wave its arms and legs and execute the most daring antics of a tightrope walker that anyone could possibly imagine. I just stood there and watched it. Suddenly, it made a great leap, extended its arms and legs, until it formed an X, and said, "decipher me or I devour thee."

…

My idea, after so many feats, became a fixed idea. God save you, my reader, from a fixed idea. Better a speck, a mote in the eye.

J.M. Machado de Assis,

Memórias Póstumas de Brás Cubas, 1881

Point of origin

What comes into your mind whenever you hear the expression "Brazilian people" or simply "Brazilians"? Outside Brazil, the answer often boils down to people dancing samba and playing football or capoeira. Or carnival: sexy women wearing G-string bikinis and feathered headdresses. Depending on how old you are, the footballers Neymar Jr., Ronaldo, or Pelé, or the actress Carmen Miranda might come to mind as well. Not by coincidence, all these figures and scenarios involve a specialized dynamic of movement or effort: hip swing. Now, why is it that a nation with over 200 million people came to be defined by the way it moves, instead of the way it looks (ethnicity) or the way it talks (language)? How did this happen? How did this unique bodily swing pervasive to physical activities like samba, capoeira, carnival, and the "Brazilian way" of playing football come to signify Brazilianness? Is it really possible to articulate identity through motion? What are the benefits of using movement as an identity? These are some of the pressing questions that this book seeks to elucidate. But before we get wrapped up in all this, let's go back to "the lady in the tutti-frutti hat" for a moment.

From all the images listed above, Carmen Miranda (1909–55) is perhaps one of the most powerful, yet controversial, symbols ever produced about Brazil. Beyond her charismatic voice and extravagant costumes, the auspiciousness with which Miranda moved to the black rhythm of samba has given international visibility to a soft and soothing way of swaying the hips. In Brazil, this swayed motion is commonly known as *ginga*. And what exactly is ginga? Ginga involves the articulation of sinuous and offbeat – or syncopated — dialogues between bodily parts, especially the hips and feet. To give the reader a glimpse of the complexity of this term, suffice it to note that people in Brazil have used many different words to describe this way of moving or the images it evokes in one's mind: *gingado* (swayed-walk),[1] *suingue* (bodily swing), *molejo* (spring, pliability, swagger), *corpo de mola* (spring-like body), *jeito or jeitinho* (way, manner, habit),[2] *jogo de cintura* (hip-play),[3] and *gambiarra* (precarious make-shift or make do),[4] to name just a few.

Here is a more interesting question: what do you, my English-speaking reader, already know about ginga? From where I stand, deeply immersed in this subject from head to toe, I no longer know what to assume. So I do what most people do these days — I *Google* it, as if I were looking up my own name on the Internet. The third result leads me to Wikipedia in English (http://en.wikipedia.org/wiki/Ginga). There I find a disambiguation page listing ten entries for the word "ginga." I quickly review them by elimination.[5] I am left with item 1, where Wikipedia informs me that ginga is "a fundamental movement of capoeira," and item 6, where ginga is described as "a Japanese-Brazilian digital TV middleware." I will come back to item 6 later on, but this book is mainly concerned with the first definition. I consider the presence of ginga in this Afro-Brazilian martial art form, along with its significance in other distinct choreographies, dance forms, and practices of everyday life.

Articulating ideas

This book derives from my doctoral research, which I developed under the mentorship of dance scholar Susan L. Foster. At first, her argument that bodies are capable of articulating ideas as *a bodily writing* (Foster, 1995, p. 15) took me off-guard. After a while, though, it "took hold of the trapeze I carry about in my brain" (Machado de Assis, 1881, epigraph). Like the character in Machado de Assis' novel, "I just stood there and watched it. Suddenly, it made a great leap..." To cut a long story short, her idea moved me to design a comparative study to measure and qualify how ginga works, what it means, and how it interacts with other systems of representation. Part 1 of this book is entirely designed to fully explore this. For now, here is what you need to know. First, *ginga* means the movement you see when a man sways his hips or a woman shuffles her feet. Secondly, this particular way of swaying and shuffling is a key element within a movement system anchored in Africanist aesthetic principles. Polycentrism and polyrhythm are two of them. *The ginga aesthetic* is, by distinction, the expression I use to talk about this movement system. It is the underlying map or structuring logic that guides or enables bodies to move in such a fashion.

One of the strengths of this project is a set of analytical tools (i.e. lenses, approaches, and apparatuses) that I have compiled and tailored from various disciplines. I use some of them to analyze how this movement system works and how it is used in distinct realms. In this book I pay close attention to the presence of ginga across three specific and distinct realms: samba, capoeira, and concert dance. In the last case, I analyze the incorporation of the ginga aesthetic in set choreographies by Grupo Corpo, a contemporary dance company based in Belo Horizonte, Minas Gerais. In the end, rather than providing a comprehensive examination of these realms in Brazil, in each case my movement analysis focuses primarily on three aspects: (1) how ginga *functions* at the bodily level; (2) the multiple *roles* it assumes in each of these realms; and (3) the *effects* that it mobilizes through choreographed and/or improvised movements, trajectories, or knowhow.

Across the chapters, my genealogy of ginga observes how a bodily mechanism connected to Afro-Brazilian heritage articulates blackness and how it interacts with other categories of identification (e.g. gender, sexuality). Some of the tools below have helped me examine, in particular, how a slippery signifier is first read as a symptom of primitivism and/or immorality, but eventually becomes a floating signifier for that which is local/national. Departing from Spinoza's inquiries regarding human affect (*affectus* in Latin), I examine how the deployment of this movement system has engendered various actions and passions (e.g. desire, pride, shame, and melancholia) at the bodily level. Like Spinoza, I am particularly concerned with the means through which these affects increase or diminish one's power of acting and/or thinking (Spinoza, 1993 [1677], pp. 68–112).

Where I am coming from and where we are going

In a nutshell, this comparative study analyzes the contributions of Africans and their descendants across multiple and distinct spheres of Brazilian culture. It dialogues, first and foremost, with the domain of critical dance studies, which advocates for a contextualized specificity of its analytical tools. It also expands its field of vision beyond Europe, US, and other geo-political locations where English acts as a "lingua franca" (i.e. the former British and US empires; see O'Shea, 2010). When I began my research in 2005, for instance, "critical dance studies" research and projects were being funded and tested out largely within North American and European institutions. Often, these investigations privileged proscenium stage productions over vernacular and/or non-Western movement forms. These other practices and forms were still being examined, with few exceptions, within the fields of cultural studies, anthropology, sociology, and folklore (Savigliano, 2011).

This had also been the case in Brazil. Movement practices associated with Afro-Brazilian heritage had been recorded and analyzed in Brazil for more than a century.[6] When compared to the scientific racism of the nineteenth century, pioneering works such as Cecília Meireles' *Batuque, Samba e Macumba* (1933, an ethnographic exhibition of drawings and watercolors later published in 1983) and Mário de Andrade's ethnographic catalogue of Brazilian folk dances and pageants (*Danças Dramaticas do Brasil*, originally published in 1934) represent a remarkable turning point in terms of expanding the meaning of culture to include folk and popular manifestations. In the end, though, these earlier authors had little experience of dancing or writing about dance. They had no way of giving the reader a clear sense of how, for instance, the female dancers of the batuque, samba, and macumba moved their hips underneath their skirts. At best, these publications describe the footsteps that these dancing bodies left behind. The reader was often left to believe that the dancers did it "naturally."

The recent scholarship on Afro-Brazilian social dances such as samba (and its relation to carnival) has injected the field of dance studies with a number of fertile seeds.[7] In the last decade, publications such as the National Institute of Historic and Artistic Heritage's dossiers on intangible heritages have also expanded and systematized the scholarship on these subjects.[8] However, when I began my research, little work had been done to address how ginga informs or relates to samba dancing. Similarly, in the 1990s, there was a revitalization of the field of dance studies in Brazilian universities and its scholarship is progressively solidifying both its theoretical and practical elements. In particular, these institutions have sought to diversify the sources and outputs of their productions. But this has been a gradual process.

In the case of capoeira, there has been a substantial proliferation of ethnographic and historical research projects about this martial art in the last 20 years. Most of these works seek to address several, if not all, of capoeira's structural and performing aspects.[9] Generally speaking, scholars have cross-referenced a wide range of archives against their observations of particular performances (repertoire) and/or their fieldwork (ethnographic material). Yet, like those writing about samba and carnival, most lack the training and the necessary tools to analyze capoeira's non-verbal discourses. Instead, these scholars tend to focus largely on knowledges transmitted through oral history, artifacts, musical rhythms, and lyrics of songs.

In order to overcome these multiple gaps and shortcomings, I reached out to distinct strands across the field of critical dance studies. I selected and mixed theoretical tools pertaining to dance ethnography, historiography, and close analysis of set dances and movement practices.[10] I bent them and re-shaped them, when necessary, to make them work in my project. I also pulled additional tools from other disciplines, such as art history, performance studies, critical studies of race and gender, phenomenology, semiotics, and recent developments in post-colonial studies in Latin America. All along, I have made a conscious effort to slide back and forth across these various fields, at times overlaying multiple concepts at once, and at other times executing all kinds of acrobatic maneuvers. Below, I briefly acknowledge some of the most significant elements in my "tool box" and their deployment and relevance.

The expression "choreographies of identification" is an amalgam that intersects a number of propositions. For starters, it follows the understanding of bodies as both the producers and the medium of their own productions. For dance scholars Helena Katz (2003, 2005, 2006) and Christine Greiner (2005, 2010),[11] bodies and environments are constantly exchanging data, or shaping and being shaped by one another, through continuous processes of mutual *contamination*. Here, it should be noted that the term "contamination" acts as a provocation against discourses and practices that either defend an imagined state of ethno-cultural "purity" or prescribe racial "hygiene."[12] Named *corpomedia*, their theoretical lens provides a collection of anchor points from which to discuss the epistemology of the body and cognitive processes in dance. With this tool at hand, I consider the ways in which the ginga aesthetic informs how people organize their bodies to articulate ideas in scenarios such as concert dance, social dance, and martial arts. It also allows me to consider the space people construct as they move around and how the environments where they step into also shape these moving bodies (see Part I).

This amalgam also dialogues with the phenomenologist and media theory scholar Vilém Flusser (1998, 2014). In the 1970s, Flusser began to write a manuscript for what he called a "General Theory of Gestures" (see Flusser, 2014).[13] His writings on gesture turned my attention towards the role of physicality in discussions of knowledge production and ways of knowing. His philosophical discussion on bodily cognition made me consider the relevance of one's relationship to one's body as well as the underlying logic through which one thinks and acts within these movement practices.

The black Atlantic world (Gilroy, 1993) to which we are about to sail is a place filled with a transnational, multi-lingual, and multi-religious flux of people and ideas from different ports. This amalgam therefore departs from the understanding of categories of social identity such as "black" as intersectional and pliable *processes* situated in a particular geopolitical context or dimension. As the critical race theory scholar Mara Viveros Vigoya clarifies in her essay "Dionysian Blacks: Sexuality, Body, and Racial Order in Colombia":

> Here "black" is understood not as an essentialized identity but as a personal, social, cultural, political, and economic process in a particular temporal and spatial context with local, regional, national, and trans-national dimensions. (Viveros, 2002, p. 75)

When I examine performances centered on ginga, I assume that they have been shaped by heterogeneous set of ideas under which African peoples and their descendants have (re)structured their communities, their systems of beliefs, and their systems of knowledge production in Brazil. With this understanding at hand, I problematize a number of essential assumptions, such as: (a) the belief that blacks are inherently inclined (i.e. feel an impulse) to seek sexual pleasure; (b) the limiting equation of racial miscegenation to processes of whitening; and (c) the belief that the essentialization of one's identity is the only way to valorize alterity (e.g. black pride).

Following this strand of thought, I pulled the expression "processes of identification" from the writings of Jesus Martín-Barbero (1987, 2003). I use it here to differentiate "essentialized identities" (i.e. *fixed* ideas, images, and discourses whose variables are imagined static and constant) from *flexible* and complex processes articulated across a range of intersecting categories (gender, race/ethnicity, sexuality, class, nationality, etc.) situated in a particular time and space. I also dialogue with Amalio Pinheiro's understanding of individuals as "movable territories" (2008). Like Katz and Greiner, Pinheiro defends the idea that people are constantly interacting with strands of information coming from different places.

Most prominently, I have adopted Susan L. Foster's definition of choreography as an overarching score or a "framework of decisions that implements a set of representational strategies," which "evidences a theory of embodiment" (1998, pp. 16–17).[14] Employing choreography as a theoretical lens, my comparative study investigates the ways in which the ginga aesthetic interacts with choreographies of gender, ethnicity, sexuality, and, later on, nationality. Foster's deployment of choreography has also enabled me to examine how bodies (are expected to) look like, move, behave, and interact with one another and, thus, how choreography relates to other systems of representations (Foster, 1998). In her recent scholarship, Foster points to "the existence of corporeal epistemes that participate in the production of knowledge and the structuring of power" (2010, p. 13). Applied to my comparative study, this framework allows me to see more clearly how the cultural interactions in a country like Brazil happen at the bodily level. Departing from this premise, I consider how the colonial encounter in Brazil generates not only the (forced) interaction, or at least coexistence, of diverse peoples, languages, and cultural goods, but also the transculturation (Ortiz, 1995) – or rather the *recuperation-cum-invention*[15] – of a range of ideas informed by heterogeneous notions of corporeality, connected to various corporeal epistemes.

In the end, the amalgam "choreographies of identification" intends to contribute to the theorization of identity formation as a pliable, decentralized, and multilayered process. In order to do this, I combine (a) Foster's understanding of choreography, corporeality, and corporeal episteme as elements implicated in the production of aesthetic knowledges and discourses of identities (i.e. gender) with (b) Katz and Greiner's view of bodies as both agents and media of themselves, (c) Flusser's recognition of the phenomenological relationship between moving and thinking and its relevance to human cognition, (d) Martín-Barbero's definition of processes of identification, and (e) Pinheiro's understanding of individuals as movable territories.

This book offers a comparative analysis of diverse practices whose assimilation of the ginga aesthetic has yielded different results. When examining how bodies and environments exchange information in movement practices, my comparative study benefits from Diana Taylor's (2003) framing of performance as an interpretative lens with which to examine the colonial encounter in Latin America as well as her use of the concepts such as scenario, acts of transfer, linguistic tropes, systems of equivalences, and double codedness. Taylor's scholarship pays close attention to how traumatic memory percolates through Latin American productions. By contrast, I second Margaret Drewal's (1992) and Rita Amaral's (2002) understanding that, across the African diaspora, playfulness is a rhetorical strategy that is pervasive to various events and celebrations. It thus informs the way in which these communities transmit knowledge and make sense of the world. As will be addressed in Chapter 1, along with coolness and dissonance, derision and serious play function as generative forces in the emergence of transatlantic cultures and identities in Brazil. In the context of religious rituals of Candomblé, Amaral (2002) points out that the labor-intensive and body-centered celebrations encourage the serious dissipation of physical energy in dancing, singing, and playing music, and the replenishing the body with votive food, beverages, and prayers. At these religious parties, the lavish consumption and expenditure of resources enable priests and devotees to maintain their epistemology alive.

In order to measure and qualify the elements associated with the ginga aesthetic, in my research I pulled ideas from scholars addressing African/Africanist aesthetic knowledges in movement, especially Thompson (1966, 1973, 1983), Tavares (1984), M. Drewal (1992), Gottschild (1998), and H. Drewal (1999), fully discussed in Chapter 1. My methodology also takes into consideration how blackness is articulated in both the religious and vernacular spheres.[16] I embrace, for instance, Tricia Rose's assertion that the innovative aesthetic of urban practices in black America (i.e. hip hop) has the potential to incite "social dislocation and rupture" (1996, p. 196). In my comparative study, I look for instances where people have questioned hegemonic modes of identification, producing movement styles "no one can deal with" (Rose, 1996). In order to investigate the ways in which ginga choreographs blackness in post-colonial Brazil, I also dialogue with Thomas DeFrantz's notion of "corporeal orature" (2004). In addition,

Yvonne Daniel's (2005) work on the African diaspora, especially her corporeal cartography and her understanding of bodies as "active repositories," has inspired me to imagine the ginga aesthetic as a mapping system — an underlying set of cultural codes in which bodies articulate images and metaphors. Throughout this book, I deploy this mapping (see Figure 1.1 in Chapter 1) to decipher the embodied principles it departs from and the ideas it (re)produces through set and improvised choreographies. I watch for patterns when comparing what happens in each distinct realm.

This book defends the socio-political significance of movement as a non-verbal way of knowing and articulating ideas corporeally. It further looks at how the migration of embodied ideas to different scenarios, and their incorporation by different kinds of bodies, transforms perceptions of culture, identification, representation, and self-representation. Following Marta Savigliano's (1995) steps, I have borrowed the "tools of the colonizer" (i.e. feminist, post-colonial, and post-structuralist discursive frameworks) to scrutinize the formation of transatlantic choreographies of identification in Brazil. Savigliano's detailed unpacking of a process of auto-exoticization in a global capitalist economy enabled me to comprehend the underpinnings of Latin American hybrid dances and the kinds of power relations they negotiate on the dance floor. In doing so, this book opens up new paths to dialogue with the field of Latin American studies, offering the possibility of considering how people put meaning in motion (Desmond, 1997). When compared with Latin American scholarship on literature, theater, music, visual arts, and oral history, the understanding of movement as a primary source of information unfolds a new range of possibilities.

I also rounded up some tools advocated in the work of scholars looking at dancing endeavors in Central America and the Caribbean, and their relevance to Latino culture in the US. Noteworthy are: (a) Alicia Arrizón's (2002) deployment of Ortiz's concept of transculturation to problematize the subaltern agency of mulata dancing bodies and the racialized performativity of their representation in the mass media; (b) Melissa Blanco Borelli's (2008) coinage of the term "hip(g)nosis" to problematize the "tragic mulata" and (re)locate her source of empowerment in her choreographed actions; (c) Pricilla Ovalle's (2010) demonstration of how Jennifer Lopez's "urban *sensualidad*" choreographs ambiguity in her negotiations of power relations; and (d) Cindy García's (2013) ethnographic discussion of the socio-economic trajectories of salsa dancers (*salseras*), their homosociality, and the corporealities they articulate on the dance floor.

The ethnographic component of this interdisciplinary study (i.e. samba and capoeira) has been greatly informed by Sally Ness' (1992) insistence on taking into consideration the socio-political context of cultural events and Cynthia Novack's (1990) approach towards critically investigating dance practices and performances as culture. I adopted, rather methodically, Novack's close attention to "the social and public nature of dance and the ways in which dance creates meaning" (1990, p. 13). Based on the work of Laban and Bartenieff, Novack's methodology has enabled me to analyze improvised choreographies up close. In Chapter 4, I employ Novack's analytical tools to examine the functions and roles of the ginga aesthetic in capoeira angola's call-and-response games, as well as the effects that its deployment produce. I also consider the effects of flexible choreographies of identification, which elicit both self-assertion and dignity. When examining the ways in which Grupo Corpo has appropriated the ginga aesthetic in its choreographies, in Chapters 6 and 7, I weave in Foster's (1988) methodology for reading dancing with DeFrantz's (2004) close reading of Alvin Ailey's choreographies. Choreographing sassy hip swings onto local ballet-trained bodies, for instance, the post-colonial mixing experiments of Grupo Corpo have contributed to blur the "abyssal line" (Santos, 2007) constructed between two systems of organization (e.g. the ginga aesthetic and ballet technique).

I have also pulled concepts from the recent work of a group of post-colonial scholars in Latin America commonly identified as the "Coloniality of Power" group.[17] My comparative study seeks to plant their ideas in the field of critical dance studies, opening up a dialogue regarding the recognition of moving bodies as producers of situated knowledge. Besides, these scholars' understanding of how the colonial matrix of power continues to produce effects (including bodily affects), even after the end of colonialism (Quijano and Wallerstein, 1992; Quijano and Ennis, 2000; and Mignolo, 2002) offers a productive way of critically analyzing the conditions under which subjugated bodies have deployed movement in their processes of self-fashioning. In that sense, this book sheds new light on renewed processes of resistance, especially the connection of dancing and dance making to different economic, socio-political, and cultural systems and procedures.

Through the combined use of these diverse tools, ginga is taken here as an entry point for a set of socio-historical and transcultural reflections regarding embodiment, knowledge production, and processes of identification. In my comparative research I have reached out to the field of critical race studies and, in return, this book generates valuable information on: (a) how Africans and their descendants have shaped communities in the New World; (b) how embodied knowledges connected to African heritage (e.g. polycentrism, polyrhythm, call-and-response) have helped to decolonize as well as to recolonize individuals and communities in the diaspora; and, subsequently; (c) how non-hegemonic movement systems (i.e. ways of organizing bodies to think and act) may function as tools against the epistemic violence that resulted from colonization. In the case of Brazil, where African descendants constitute the majority of the population, new embodied identities have emerged through an active process of selection of movements, gestures, and intentions to signify blackness. This complex process has triggered the (re)invention of set and improvised choreographies about themselves, their imagined communities, and the world in which they live.

As a result of this multi-faceted dialogue, my analysis of the presence of ginga across daily practices, social dance, martial arts, and concert dance demonstrates how ideas take new shapes and how forms acquire new meanings as people move around. As the following chapters seek to demonstrate, individuals and communities are continually (re) constructing their contextualized processes of identification through lexicons of movement that they watch, applaud, learn, rehearse, select, forget, reject, transfer, recuperate, revisit, reinvent, talk about, and perform for and about themselves and about others. In addition, set or improvised choreographies, sequences, or even qualities of movement performed repeatedly also transform people's perception of the world, in as much as they rearrange how their bodies are "wired" kinesthetically to think and act. In other words, the data connected to any collection of moving ideas, both the underlying system supporting their format and the knowledge to structure the body to perform them, is transmitted, stored, and reproduced physically. Over time, the embodied process of production and exchange of information through movement in a given scenario or context leads to lasting transformations within the way bodies mix new and old resources. Now, since both these bodily technologies and their underlying systems of organization shape the moving bodies into particular forms and relations, they subsequently transform their ability to interact with and understand the world around them. Ultimately, these interactions between moving-thinking bodies from different walks of life and choreographed ideas from distinct ethnocultural backgrounds provoke complex processes of mutual "contamination" between them and in response to their socio-historical context. Yet, choreographies of identification and their underlying systems of organization are not "infectious" or "contagious"[18] in the sense that they will rarely alter bodies after brief exposure. It is the continuous practice (or rejection) of a particular set of embodied principles, i.e. the painstaking process of learning to articulate them and the construction of the repertoire of one's life around these ideas, which eventually molds one's sense of selfhood. Likewise, the continuous rehearsal and re-articulation of specific embodied knowledges in social dance, martial arts, theatrical dance, and music, as well as their representation in visual arts, architecture, city planning, cinema, or on TV, which eventually molds what a community becomes. The way in which people choose or are forced to use their bodies also transforms their perception, description, and reflection over the tangible world. Conversely, ideas that people are able to see, hear, touch, smell, feel (emotionally), and conceive (intellectually) about the world around them, or take for granted, mold the way in which they experience, imagine, and orient themselves, as well as how they communicate with others.

One last point requires attention. In an ambitious quest to combine distinct tools and approaches (e.g. historiography, movement analysis, ethnography, and phenomenology), mixing theory and practice across multiple sites, in and out of Brazil, I have also assumed multiple positions. When writing about Grupo Corpo or Carmen Miranda, for instance, I have observed my object of study at a safe distance, diving into archives, watching DVDs, attending performances and dance rehearsals, or talking to people on the street. Nevertheless, when writing about capoeira and samba, my object of inquiry often came into close proximity with me and, on multiple occasions, the author became the central focus of her own inquiries, thus blurring irreversibly all the distinctions between subject and object. As a result, the author of this book has not only learned about ginga aesthetic, in all its varied manifestations and facets, but she has also learned how to move that way. From a distance (i.e. talking about myself in the third person), *her* way of thinking and acting have become distinctively "contaminated" with the ginga aesthetic. Following this heterogeneous logic, the organization of the book you are reading tends to sway from side to side. This is not an excuse, but rather an acknowledgement of the rhythm and the flow of my remarks. At any rate, it is worth warning the reader that, for better or worse, the body writing this book tries to expose what the ginga aesthetic *feels like* to those reading about it. Inadvertently, the argument slides away from its expected linearity. Hopefully this won't cause you, my reader, to feel dizzy or faint.

Brazil's pride-and-shame conundrum

Now let's switch gears and look at what is at stake when one wiggles the hips or thinks of ginga. On the one hand, since colonial times, Africans transplanted to the New World and their descendants have employed a polycentric and polyrhythmic way of moving to negotiate renewed intersubjectivities, from a place of otherness. In Brazil, they cultivated a range of syncopated dances, games, and rituals within secluded black spaces (i.e. slave quarters and maroon communities). In addition, they have also appropriated the public (festive) space and Western/colonial theatrical devices to enact flexible choreographies of (self-)identification centered on the ginga aesthetic. They improvised makeshifts and gambiarras of all kinds, mixing tradition with hyperbolic imagination, borrowing elements as necessary. In all these cases, the recuperation-cum-invention of Africanist aesthetic principles in the New World has contributed to empowering disenfranchised communities and restoring their sense of cultural pride, moral self-esteem, and social dignity.

On the other hand, one should also note that the care of self, exercised across practices centered on the ginga aesthetic – especially spine undulations and hip thrusts – clearly opposed the (normative) asceticism of the Western/Christian culture of the self. Subsequently, within the colonial archives, ginga has invariably been recognized as an uncivilized or dishonest behavior, indicative of the racial inferiority and/or the moral degeneration of its practitioners. In fact, prior to the Second World War, the European-aspiring elites in Brazil continued to perceive movement of bodily syncopation largely as the cause or symptom of social/moral diseases. Ginga was understood, from this point of view, as a despicable or shameful act. This interpretation has contributed to fix individuals engaged in Afro-Brazilian practices such as samba and capoeira as not honorable or not trustworthy subjects.

To summarize, the coexistence of these antagonistic understandings of ginga in Brazil and the different affects it produces (e.g. pride, shame) has generated a set of complex (and unresolved) ideas that I recognize throughout this book as a *pride-and-shame conundrum*. As I will dissect in detail in Chapter 2, this conundrum engenders a number of negotiations connecting the ginga aesthetic to Brazil's process of identity formation. More importantly, this set of complex ideas propels from and signals to a paradoxical yet deep-rooted set of relations particular to that nation's patriarchic colonization: the sexual/cultural interaction between races (miscegenation) inside a country socio-economically built on the racial division of labor (slavery) and Eurocentrism. As I will demonstrate, Brazil's pride-and-shame conundrum remains alive, even after the abolition of slavery (1888). In particular, the ruling classes' perception of miscegenation as a threat to development and progress of that society provoked a kind of national "identity crisis" during the First Republic (1889–1930).[19] Nevertheless, in the 1930s, Brazilian society embraced a new discursive antidote that, similar to "the poultice" that Machado de Assis' character invented, his "acrobatic idea,"[20] was powerful enough to attenuate the nation's melancholia. Below, I describe this sublime remedy, outlining its uses and side-effects.

The poultice

The current notion of Brazil as a modern nation or imagined community (Anderson, 1991) emerged in the first half of the twentieth century. Moving away from scientific racism, in the 1930s, the sociologist Gilberto Freyre departed from Franz Boas' notion of cultural relativism to propose at a revolutionary understanding of cultural miscegenation as a positive element of Brazil's uniqueness.[21] Meanwhile, the ambiguity of Freyre's idealized mulatto culture (*mulatismo*) perpetuates, or at least extends, the reoccurring avoidance to discuss the role of blackness (*negritude*) on the national arena (Carvalho, 1996; Nascimento, 2002). Giacomini (1988) points out, furthermore, that Freyre's theory ignores the physical and the epistemic violence of the colonial encounter, which invariably tainted the dignity of black female bodies with shame.

Despite all that, Freyre's sociological interpretation of Brazil took that nation by storm, provoking a radical transformation in the way in which Brazilians, especially the intellectual elites, saw themselves. More importantly, his decision to write about popular culture practices disseminated via means of mass communication (e.g. samba music and football) amplified the popularization of his ideas. During the 1938 World Cup in France, in which Brazil achieved a glorious third place, Freyre published a newspaper article entitled "Foot-ball Mulato" (1938), in which he considers the concept of "cultural mulatism" in that sport. In the article, Freyre argues that local footballers in Brazil incorporated elements from Afro-Brazilian practices (i.e. samba and capoeira) into their way of playing that European sport. As he proposes, Brazilians transformed the British way of playing, or "Apollonian" style, into a "Dionysian dance" (1938). Later on, in the preface of Mário Rodrigues Filho's quintessential *O Negro no Futebol Brasileiro* (*The Black in the Brazilian Football*), Freyre makes his viewpoint more transparent. As the country geared up to host the next World Cup in 1950,[22] Freyre (1947) affirms that the importance of football in Brazil lies in its ability to sublimate animal energies and irrational instincts present in all Brazilians, but especially so in blacks and mixed-race footballers. Implicit here is the (Eurocentric) idea that the practice of the British "sport" could transform non-white/primitive individuals into civilized *sportsmen*.

By the end of the 1930s, the singer/actress Carmen Miranda (discussed in Chapter 5) and the Afro-Brazilian footballer Leônidas da Silva were amongst the most famous names in the entertainment industry in Brazil. More importantly, their performances fit Freyre's concept of cultural mulatism like a glove. Unlike Carmen Miranda, a figure who looked white but acted black, Leônidas da Silva gained notoriety for his ability to excel in playing a European sport or, rather, "acting white." In the 1938 World Cup in France, for instance, Leônidas da Silva received the top scorer award of the tournament. But like the "lady in the tutti-frutti hat," his uncanny maneuvers were exoticized on both the local and global stages. In 1938, for instance, the French press compared Leônidas da Silva to a circus freak with six legs and an "elastic man" (Ribeiro, 1999).

Meanwhile, in 1930 President Getúlio Vargas rose to power and instituted a paternalist posture that sought to get a handle on the question of national identity, which had been unresolved until that point.[23] During his populist dictatorship, or New State (*Estado Novo*, 1937–45),[24] in particular, we see the emergence of a new regime of intelligibility, today commonly referred to as the "democracy of races" ideology. Departing from the ambiguity of Freyre's concept, amongst others, this ideology re-packages cultural miscegenation as a positive outcome of colonization. In the end, the political economy of indefiniteness inherent in Freyre's pliable understanding of mulatismo contributed to supporting the ideological conceptualization of Brazil as a racially harmonious community during the Estado Novo. However, this "sublime remedy" certainly did not resolve the question of race/racism in the country. In sum, the "democracy of races" ideology converges a positive light towards "mixtures" unique to Brazil (e.g. mixed bodies and mixed culture), but it does not question the colonial/patriarchal "mixing" procedures and motivations. Like the colonial division of labor that justified slavery, this modern "poultice" became a fixed idea (see Machado de Assis, 1881). As I will argue across this book, in forms such as samba (and football), and later on capoeira and concert dance, this regime of intelligibility finds the necessary conditions for its representation, validation, and mass dissemination.

Little by little, the carefully constructed process of imagining Brazil as a racially harmonious community across the twentieth century has contributed to *fixing* this once marginalized way of moving as a central symbol of Brazilianness. Meanwhile, the racial division of labor, its resources and products, remained unquestioned. In other words, the ingredients put into the "Brazilian" mix continue to be stored in hierarchic and discriminating places where whiteness is (essentially) "good," while blackness is (ontologically) "bad." After the Second World War the value allocated to the ginga aesthetic began to oscillate back and forth within mainstream society inasmuch as it evoked blackness, was tamed or framed by whiteness, or found a dynamic (indefinite) balance between the two.

Ultimately, by combining historiography and ethnography with choreographic analyses, this book offers an insight into post-colonial processes of identity formation in Brazil. Today ginga is at the heart of the image that people across the globe have of Brazilians. On national soil, nevertheless, ginga continues to incite contradictory affects, from pride to shame, depending on who articulates it, and when, where, how, and why it is performed, represented, or insinuated. More often than not, though, this way of moving has been incorporated into fixed (and limiting) amalgamations, where gender and sexual identities are represented according to a marked race. As I will demonstrate, these fixed choreographies of identification continue to generate auto-exotic figures (e.g. the mulata who dances samba or the dandy who plays capoeira) that, although feared or desired, are seldom regarded in an honorable or dignified matter.

The argument of this book is threefold. I first argue that the presence of ginga in non-verbal discourses currently articulated in Brazil provides us with wide and concrete evidence of Africans and their descendants as producers of knowledge and ways of knowing. It also argues for their active role in the modern construction of Brazil as an imagined community. Secondly, I propose that *the system of bodily organization and knowledge production* cultivated in Afro-Brazilian movement practices (i.e. the ginga aesthetic), of which samba and capoeira are the most widely known, have contributed to *recuperate-cum-invent* an epistemology beyond colonial languages, whose scope exceeds or differs from Eurocentric thought. Thirdly, I propose that the articulation of flexible choreographies of identification centered on this movement system has led to decentralized processes of ethno-cultural resistance and emancipation, intertwining blackness with grace and dignity. Meanwhile, renewed forms of discursive control and subjugation have reduced this non-hegemonic way of organizing ideas corporeally into stereotypical symbols of Brazilianness or, rather, fixed choreographies of national identification. In this regard, the everlasting entanglement between ginga and Brazil's pride-and-shame conundrum has become a "hook that cannot be unhooked" (Savigliano, 1995, p. 75).

By way of a summary, this book addresses how the way one *moves* (i.e. choreography of identification) may function as an apparatus to negotiate power relations, in addition to one's appearance (e.g. size, shape, skin color) or social position (e.g. class, occupation, status). In my methodology, I thus focus on particular aspects (i.e. functions, roles, and effects) that this aesthetic system engenders in different realms. I note who interprets and who applauds set or improvised choreographies centered on or including the ginga aesthetic, and the context in which they are performed. When addressing the presence of the ginga aesthetic "out there in the world" (Savigliano, 2011), I pay close attention to how movement informs thinking patterns and vice versa, and how particular bodies have deployed, and continue to employ, this non-hegemonic system of organization and production as and for political action. I have examined the presence of ginga in the performances of Carmen Miranda and the repertoire of Grupo Corpo, circles of samba and capoeira, as well as daily practices such as selling food on the streets or building a house in a shantytown (*favela*). My analyses of these set and improvised scores confirm that, in the end, this ability to write and to be written, to inform and be formed by is a prerogative of all bodies, whether they occupy the position of the subject or predicate. This book therefore expands the kinds of bodies and movement practices that might offer a self-analyzing and self-reflective framework to theorize and produce signification.

In Part I of this book, I present an overview of ginga, addressing how it is embodied and conceptualized. In Chapter 1, I offer a descriptive overview of the ginga aesthetic. Briefly, this movement system pushes individuals to imagine and reorganize their moving bodies dynamically as polycentric structures in space that, subsequently, enact polyrhythmic patterns of movement in time. I then address how this movement system facilitates the articulation of other Africanist aesthetic principles, such as call-and-response, apartness, high-affect juxtaposition, serpentine pathways, dissonance, coolness, and serious play. Chapter 2 begins with an etymologic discussion of the term "ginga," contextualizing its literal and metaphoric usages across the Portuguese Empire. In addition, I discuss how the coexistence of both slavery and miscegenation in colonial Brazil pushed the articulation of ginga towards a long-lasting paradox, which I call a *pride-and-shame conundrum*. In order to illustrate this process, I historicize the presence of ginga in capoeira since colonial times.

In Part II of this book, I focus on the presence of the ginga aesthetic in three diverse scenarios. In Chapter 3, I offer a close analysis of the qualities of movement and improvisations articulated in samba circles in order to address how this movement system operates in this particular kind of social dance. By contrast, in Chapter 4, I analyze the presence of ginga in capoeira angola. Like samba circles, games of capoeira angola should be understood as an embodied way of thinking and articulating ideas (process) rather than as a cultural commodity with codified rules (product). Subsequently, I present a series of analytical tools that I have developed in order to examine the functions and roles of ginga in capoeira's call-and-response improvisations and the effects it produces. In Chapter 5, I offer a genealogy of the baiana figure, the character that made Carmen Miranda famous internationally. Moreover, I consider how samba, carnival, and this "typical" figure are fused together as inseparable symbols of Brazilianess. Subsequently, I argue that Miranda's cinematographic career in Hollywood contributes to re-signifying and calcifying samba dancing on global stages, making it look like a fixed set of rules and protocols.

Finally, in Part III of this book, I discuss the presence of ginga in the repertoire of Grupo Corpo. In Chapter 6, I discuss how nationality was represented on the concert stage across the twentieth century. I then present an overarching discussion of the presence of ginga in the repertoire of Grupo Corpo. On local stages, Grupo Corpo's ballet-trained bodies interpret choreographies centered on the ginga aesthetic, thus questioning the incompatibility imagined to exist between these two movement systems. Conversely, on global stages the company's repertoire, especially its athletic vitality and dazzling sensuality, is perceived as an authentic example of "Brazilian dance." It thus resurfaces issues similar to those raised by Carmen Miranda's cinematography. In Chapter 7, I take a closer look at the presence of ginga in two of Grupo Corpo's choreographies, *Nazareth* (1993) and *Breu* (2007). My analysis departs from the discussions of the previous chapters regarding the relationship between the ginga aesthetic and Brazil's pride-and-shame conundrum, as well as the country's development of concert dance throughout the twentieth century. Briefly, in *Nazareth* the ballet-trained bodies of Grupo Corpo employ ginga aesthetic as the main sensibility informing the entire choreography. By contrast, in *Breu* the dynamic conversations between the hips and feet break away from the soft-spoken sensuality that had been identified as one of Grupo Corpo's signatures since the 1990s.

It is worthwhile remembering that Brazil is a highly diverse and multi-ethnic nation, with a complex and multi-lateral history of immigrations and inter-migrations. As the largest country in Latin America and, in recent times, as one of the fastest growing economies in the world, Brazil's current cultural production is prolific and overwhelming. While the transatlantic paths connecting Brazil's vast territory to both Africa and Europe have not lost their vigor or level of activity, these "foundational" ties alone are not able to account for the diverse and complex movement forms historically available in the country, from rituals of passage cultivated by Amerindians living along the banks of the Amazon river to belly dancing (*dança do ventre*) and butoh (*butô*). Therefore, it is impossible to speak of a unifying "Brazilian culture," "Brazilian art," or even "Brazilian dance." Brazilians have produced many kinds of dances, arts, and cultures.

Brazil was named after a wood red as ember (*pau-brasil*) that spread out along its coast. Yet, like any other category of identification, nationality is a concept in motion, not a rooted tree. As the poet Oswald de Andrade advocates in a key text of Brazilian Modernism, his Manifesto Atropofágico (1928), Brazilians should position themselves "against plantlike elites. In connection with the soil." In his manifesto, which was later re-appropriated by Tropicália's counter-culture artists in the late 1960s, Andrade urges us to "let go" of roots and points of origin. Instead, he proposes Brazilians focus on "roteiros, roteiros, roteiros, roteiros, roteiros, roteiros, roteiros," meaning both routes and scripts.

This book invites readers to investigate how the idea of ginga has informed movement practices, choreographed actions, and thinking processes across Brazil, and why it has been desired and celebrated or rejected and criminalized at different points in time. The ginga aesthetic, this polycentric and polyrhythmic way of organizing bodies, objects, and ideas alike is a familiar pattern in Brazilian society. The ginga aesthetic is not the only system that has challenged the neutrality/normalcy of Western (embodied) epistemology; as an embodied knowledge, it is incorporated into the ways in which some, but certainly not all, people think and act in Brazil. However, over time, this structuring logic has "contaminated" different socio-cultural realms, from dance and games to music, visual arts, literature, architecture, urban planning, etc. Yet, it makes no sense to regard this movement system as the essence of the Brazilian nation or the "soul" of Brazilian bodies.

Important questions for this book are therefore as follows: why do Brazilian bodies continue to be always preconceived on global stages by fixed choreographies of identification such as sexy dancing *mulatas*, charming-yet-untrustworthy "tropical dandies" (*malandros*), or contemporary dancers whose "natural" hip-wiggling instincts and vitality dazzles the audience as they bend their technical skills out of shape? What are the socio-historical and political conditions that lead to, and continue to validate, this narrow (and rather ignorant) view of Brazil as a "Swing Nation"? How does this fixed idea work? What effects does it produce? Who benefits from it? And, more importantly, what does it hide, displace, or erase from view? What other alternative choreographies counter these ideas in exciting and new forms of dance and/or community expression?

Part I
Understanding Ginga

1
Decoding the Ginga Aesthetic

> Ginga is a way of not taking life too seriously and facing your problems with a play of hips, feet, and heels. There have been 505 years that Brazilians have been swinging [*gingando*] through life – and they can recommend it to everyone.[1]
>
> Rui Castro, *Ginga: The Soul of the Brazilian Football*

In the months leading up to the 2006 World Cup in Germany, the Nike corporation sponsored a series of worldwide parties designed to launch its new campaign "Joga Bonito" ("Play Beautiful"). In a nutshell, its goal was to connect Nike's name brand to the "Brazilian" style of playing football and its sponsored footballers (at the time, Ronaldo and Robinho). One of the key features of these marketing events was the exhibition of the documentary *Ginga: The Soul of the Brazilian Football* (2005), a partnership between Nike and the O2 Filmes production company.[2] This documentary, which chronicles the life of eight emerging and professional footballers from Brazil, asserts that ginga is an intrinsic part of Brazilian cultural identity. Juxtaposing scenes of Brazilians dancing samba, playing capoeira, and playing soccer, the irreverent documentary revisits Gilberto Freyre's 1930s argument that the bodily swing – i.e. the ginga – present in Afro-Brazilian practices such as samba and capoeira have "contaminated" that European recreation in Brazil, transforming the Apollonian sport into a Dionysian dance or "foot-ball mulato" (see the Introduction).

On the back cover of the box set distributed during Nike's campaign, more importantly, one reads that ginga is:

> an almost indefinable, mystical quality of movement and attitude possessed only by Brazilians and evident in everything they do. The way they walk, talk, dance and approach everything in their lives. Ginga is what gives Brazilian football players their fluidity and rhythm on the pitch and enables them to "Joga Bonito" (Play Beautiful).

Contrary to this view of ginga as "an almost indefinable, mystical quality of movement and attitude possessed only by Brazilians," in this chapter, I will argue that ginga is a bodily disposition or way of moving that, though connected to a particular culture, might be acquired, embodied, and reproduced by anybody, in spite of his or her ethnicity or place of birth. Subsequently, in this book I will offer an analytical discussion of the underlying principles informing the embodied concept of ginga as well as an introductory discussion of how this bodily mechanism works in practice. Ginga may be defined as a swaggering way of sliding or tilting (parts of) the body from one side to another when walking or, otherwise, acting in society. It functions, more importantly, as a central mechanism with which one may "juggle" weight across time and space, while maintaining a cool and supple sense of flow.

As stated in the Introduction, this "juggling mechanism" is the key element of a non-hegemonic *system of bodily organization and knowledge production*, which exercises a distinct way of perceiving and interacting with others. For practical reasons, hereafter I refer to this movement system as the *ginga aesthetic*. Based on the information discussed on this chapter, in this book I draw attention to how different segments of society in Brazil have cultivated steps, gestures, and intentions centered on the ginga aesthetic, selectively remembering and recombining them with other lexicons of movements available in, but not limited to, indigenous and foreign dance forms, martial arts, and team sports.

Preliminary considerations

For starters, it is important to recognize that there is no scientific evidence that Brazilians, or even Afro-Brazilians, inherit dispositions such as bodily syncopation genetically. In this regard, the music composer and literature professor José Miguel Wisnik states that:

> In order to escape the risk of idealizing Brazil and Brazilians, we need, from the very start, to refrain from postulating cultural superiorities,

> or even from positing an idea of "national character." Instead ... it's important to identify complex specificities. (Wisnik, 2011, p. 3)

For Wisnik, for instance, the complex interactions of African and European peoples and cultures in the New World, especially in major pro-slave port cities such as Rio de Janeiro and Salvador in Brazil, New Orleans in the US, and Havana in Cuba, have led to distinct, yet comparable, processes of "Africanizing transformation" and "hybridization" of cultural spheres. In relation to the "Africanization" of music in the New World, for instance, Wisnik further points out that:

> The European polka, which is the prototype of all popular, urban dance music, becomes *ragtime* in the United States, *habanera* in the Caribbean, and *maxixe* in Brazil. The rhythmics [*rítmica*] of these three genres from North, Central, and South America are equally *contrametric* [*contramétrica*]: that is, they base themselves on accentuations that fall outside of the *tonic points* [*pontos tônicos*] of the binary measure; they thereby create a texture of internal pulses, tending towards polyrhythm, which call for a swaying dance [*dança gingada*] full of swing. These rhythmics are different from the ones that predominate in Western Europe, which musicology identifies as *cometric* [*cométrica*]: that is, where the figures and rhythmic divisions double the *tonic points* of the measure. (Wisnik, 2011, p. 3, italics and brackets in original)

Drawing on Sandroni (2001),[3] Wisnik explains that this frictional interplay of rhythmic *cometricity* (same meter) and *contrarmetricity* (different meter or multi-meter) infuses the main rhythm with different kinds of polyrhythmic patterns or pulses, identified in Western music theory as *syncope*. This explains why samba, rumba, and jazz rhythms have "swing." In a related article, Wisnik further explains that syncope may be understood as "a constant oscillation between an order and its accentual counter-order, sustained in a single movement" (2004, p. 46, author's translation) or rather a stress in non-tonic points within the basic riff. Therefore, it presupposes pliable dislocations and discrepancies. Moreover, "this kind of rhythmic ambivalence, or structuring oscillation within the tonic points of reference, demand of the active listener a balancing sway, a characteristic and counter metric swing of the body" (2004, p. 48, author's translation). The syncope installs, he adds, a dialectic between two simultaneous measures of accentuation, which the Brazilian Afro-European rhythm sustains to the limit: the binary beat that its counter-measure pulls, and the combined addition of odd and even musical cells that agglutinate and subdivide themselves, within the main riff. I will return to Wisnik's understanding of syncope when discussing qualities of movement related to ginga.

Below I offer a brief overview of the Africanist aesthetic principles informing this way of moving. My theoretical conceptualization of ginga as the foundational element within a movement system that *recuperates-cum-invents* an Africanist way of thinking and acting, or embodied epistemology, draws on the groundbreaking work of the art historian Robert Farris Thompson on West African aesthetics and its presence in the diaspora. For Thompson (1966), the five characteristics shared by a wide range of dance forms in West Africa, and their respective music traditions, include *apart playing and dancing*; *dominance of the percussive concept of performance*; *multiple meter*; *call-and-response*; and *songs and dance of derision*. Together, they amount to what he identifies as the West African "aesthetic of the cool." Over time, Thompson's considerations on the aesthetics of West African dance and its cultivation on both sides of the Atlantic have triggered a radical change in the course of scholarly studies across the black Atlantic world. In particular, the findings of his multi-sited ethnographies have provided researchers and practitioners with concrete anchor points from which to identify, trace, and critically analyze how Africans and their descendants contributed to the reproduction of aesthetic knowledge in the Americas through movement.

Dance scholar Brenda Dixon Gottschild (1998), for instance, has followed in the footsteps of Thompson, discussing the premises of an Africanist aesthetic derived from West African principles and its identifiable presence in a variety of dance forms and regimens of training developed in North America. Moving one step further, Gottschild acknowledges the direct relationship between meaning and motion, inferring that dancing bodies articulate ideas and worldviews by deploying aesthetic principles through actions and attitudes. Her analysis addresses, for instance, the position and the articulations of the torso, the relationship between the torso and the limbs, and the resonance between rhythm and movement.[4] Similarly, in his comparative ethnography, the sociologist Julio Tavares (1998) has traced a correspondence between the concept of coolness (related to jazz), as observed in Harlem, New York, and the concept of ginga (related to capoeira), as observed in Mangueira, Rio de Janeiro. Connecting embodied knowledges to processes of identification, Tavares concludes that in the diaspora, men and women's signature ways of moving construct renewed self-representational images of elegance and beauty (coolness in the US and ginga in Brazil), whose performativity is further amplified by bodily modifications such as hairstyles, accents, clothing, and swagger.

The cross-examination of the available scholarship on African/Africanist aesthetics and its proliferation in the New World measured against my embodied knowledge of particular Afro-Brazilian movement practices grounded in the ginga aesthetic (e.g. samba, capoeira, frevo, and coco) has led me to outline a series of principles that are crucial to the understanding of the ginga aesthetic, and therefore the argument in this book. For the sake of clarification, I have grouped principles that are mentioned in this research project under eight subheadings, some of which overlap with one another and some of which are not fully developed. Yet, in this book, a clear understanding of the principles listed below becomes handy when retracing and qualifying the presence of this underlying system in a variety of socio-cultural activities, from dancing to thinking. The following lines function as a glossary for the text.

First and foremost, the concept of *polycentrism/polyrhythm*, as defined by Gottschild (1998), refers to movements generated from more than one body part and/or connected to multiple meters. By extension, the *polycentric and polyrhythmic body* alludes to the fragmentation of the body into multiple centers capable of articulating more than one rhythm simultaneously. With this infrastructure in place, dislocation in time and space may lead to two sets of variables, which I have grouped as: (a) *call-and-response* and *apartness in movement*; and (b) *serpentine pathways* and *high-affect juxtapositions*. When discussing practices such as samba and capoeira, *call-and-response* refers to dialogical conversations instantiated, back-and-forth, between body parts, between two or more dancers, between dancers and musicians, or between performers and the audience. By contrast, *apartness in movement* calls for the isolation of body parts in time and space. It also considers tangential rifts and flourishes that gain precedence over basic rhythmic patterns of movement or the search to find open spaces between the main riff with which to improvise. Therefore, it privileges an investment in (personalized) style over (standardized) form. Moving forward, Gottschild's term *high-affect juxtaposition* evokes the ability to send the body into dynamically balanced movements of contrariety that evoke irony or paradox. Juxtaposition is also observed in movements that displace, defer, disorient, surprise, or shock. It also references the abrupt or dramatic overlaps and breaks between contrasting movements or postures and the juxtaposition of body parts, as in *contrapposto*. Conversely, the ginga aesthetic also emphasizes the corporeal ability to create smooth and flexible flow that sends the body in sinuous trajectories, that is, *serpentine pathways*. Generally speaking, this principle involves undulating sways, round twists, and circular turns, as well as movements and attitudes characterized by corporeal fluidity and pliability, and other rippling effects. To "move like a snake" through sinuous dislocations in space, tangential swings, or spiral shifts in different directions is not only desired but is also utilized, in some instances, to avoid direct contact or confrontation, or to simply seduce, or confuse rather than attack the other.

While the five principles outlined above shape the lexicon of movements and the trajectory pathways associated with the ginga aesthetic, the last three inform, more widely, the flow, syntax, and rhetoric associated with this movement system. Given their importance and level of complexity, below I will describe them in more detail. Departing from Thompson's scholarship (1966, 1973, 1983), the concept of *coolness* in movement must be understood as a dynamic coordination of opposing forces. Neither too cold nor too hot, coolness is expressed through pliable and supple control, stability, composure, and equilibrium between style and character. Thompson (1983) affirms that coolness is a aesthetic manifested through idealized actions or, better, the dynamic articulation of the ideal noble character (*iwa rere*), a cold principle, in tandem with (or against) the force or energy that ignites the world (*asé*), a hot principle (Thompson, 1983).[5] Beauty is achieved, he explains, by seeking a balance between these two complementary principles (Thompson, 1973). Coolness also seeks to balance out discrete elements, such as active bodies and serene expressions, with a detached yet ignited poise. It also evokes asymmetry, looseness, and indirectness.

Departing from the concept of polyrhythm, *kinesthetic dissonance* includes stepping in and out of sync or falling out of rhythm and coming back again; sideways displacement or asymmetrical shifts; participatory discrepancies; and swinging in and out of agreement with fixed cycles (riffs). It also means to move in different, odd rhythms. These rhythmic dissonances may provoke the sense of being both connected to and detached from the main spatio-temporal organization/whole, sending rippling effects and reverberations across that shared space. Kinesthetic dissonance is expressed through a variety of movements of bodily syncopation, such as samba's frictional articulation between shuffling feet and wiggling hips or the pelvic grinding of funk carioca dancers. Moreover, in order to articulate syncopated utterances, these dancers must combine the intra-connected dialogue between decentralized and independently moving parts with their ability to reproduce multi-meter patterns, such as (tonal) syncope. Hence, a deeper look at this role of syncopation in Afro-Brazilian rhythms may help us to figure out how dancers embody the principle of kinesthetic dissonances to articulate bodily syncopation.

According to Wisnik (2004), in the particular case of samba and *maxixe*,[6] the syncope represents a rhythmic stress in a dislocated point in time; outside the binary tonal beat (2/4). However, rather than being a "mistake," tonal syncopation could be better understood as an interplay between interconnected measures, permeated by textural and asymmetrical dislocations. In this frictional dialogue between two (or more) rhythms, while some instruments may follow the same basic rhythm, others come in and out of sync by either lagging or speeding up one of the tonal intervals or changing its tonal accent. Together, they produce a constant oscillation between juxtaposed meters. Wisnik concludes that this oscillatory dissonance structures the musical score with visual and audible feints (*negaceios*), enriched with highlights and recesses (Wisnik, 2004, pp. 45–8).

Along these lines, Barbara Browning understands samba dancing as "a complex dialogue in which various parts of the body talk at the same time, and in seemingly different languages" (1995, p. 2). In a similar fashion to what happens in other rhythmic traditions of the African diaspora such as jazz, Muniz Sodré (1998) further proposes that in Afro-Brazilian dance forms such as samba, the syncope – namely the 'dragged' or 'skipped' note that alters the tempo – acts in a special manner, drawing the listener into the performance. The audience is thus 'asked' to fill in the musical lacuna or answer its 'call' with bodily actions such as vocal sounds, hand claps, mannerisms, swayed actions, or dance. In doing so, he proposes, the syncope in the musical rhythm calls "absent" or "missing" bodies to action.[7] Impromptu dancers, he insists, respond to the "dragged," i.e. syncopated, sounds and vice versa, like a call-and-response game, initially through technical and aesthetic relations, but also through affirmation of social/ethnic identification through gestures, steps, and other movement interactions. Samba dancing and its precursors stand, he concludes, as an "unequivocal demonstration of resistance against the social imperative of reduction of black bodies to a production machine and as an affirmation of the continuation of the African cultural universe" (Sodré, 1998, p. 12).

Closely connected to Thompson's understanding of dances of derision and Gottschild's conceptualization of ephebism, *serious play* places emphasis on joy, vitality, playful attack, or virtuoso agility. For Margaret Thompson Drewal, within the Yoruba world, play – *ere* (noun) and *sere* (verb) – "is an integral part of everyday life" (1992, p. 18) and a way to exchange or negotiate what you want. It points to movements, gestures, and intentions that denote a childlike state of being as well as spontaneous and "in the moment" actions such as parodies, slippages, mimesis, double entendre, trickery, and signifyin(g). Play is not, however, irrational or illogical. Unlike Western (capitalist) scenarios grounded in the dichotomy between work and play,[8] M. Drewal (1992) affirms that for West African peoples such as the Yorubas, the concept of serious play – composed of playfulness (*ere/sere*) and sacrifice/offerings (*ebo*) – reinforces improvisation, ambiguity, and slippages positively as practical knowledge; as a skillful way to maneuver in society. As will be further addressed in Chapter 4, serious play should not be mistaken for deception or cunningness, in the negative sense that these terms have acquired under Judeo-Christian morality. "Play in the Yoruba sense is an interactive exploration of the inner heads (*ori inu*) of the players, a creative, engaging, ongoing strategy for testing the stuff opponents are made of" (M. Drewal, 1992, p. 19). Serious play therefore functions as a way to intervene and/or transform reality. It may be employed as a way to defeat your enemies with a joke and even as a tactic to subvert oppression and subjugation with trickery.

Likewise, Henry Drewal (1999a) points out that across the Yoruba/Nago-influenced African diaspora, in places such as Brazil, playfulness functions as a way to accomplish tasks, to negotiate, and to have fun. As both these scholars seem to agree, within Yoruba cosmology, playing should be understood as a productive action rather than as an exclusively leisure activity. Play is subjective and might therefore have more than one meaning. In addition, in this framework, to pull a trick causes pride. In fact, within the Yoruba cosmology, playfulness and trickery are commonly associated with *Eshu* (also known as *Exú* or *Elegbara*), the messenger/trickster deity responsible for the negotiations carried between the tangible and the spiritual worlds.[9] Drawing a relationship to the African-American concept of the *Signifyin(g) Monkey*, Henry Drewal (1999) concludes that serious play engages in both cultural critique and celebration. Likewise, for Margaret Drewal: "The unpredictable trickster stationed at the crossroads, whether in Nigeria or in Brazil, is a symbol of the efficacy of play, and narratives that focus on him are models of and for its practice" (1992, p. 17). As I will demonstrate in the second part of this book, serious play has been employed in practices of resistance that successfully construct "not-so-easily-identifiable" identities, thus combating processes of objectification, through rhetorical strategies such as fakery and indefiniteness.[10]

In sum, this movement system intertwines serious play with coolness and kinesthetic dissonance, while privileging variants around a particular beat, measure, or pattern of motion. Yet, ginga does not simply mean to shake the hips or swagger with malice or cunningness. To articulate ginga, or to *gingar*, means to move with a syncopated rhythm to an individual logic that sets a person apart, while responding to a given environment or context. As is further explained below, this syncopated way of moving involves the articulation of polycentric and polyrhythmic dialogues instantiated across different bodily parts, especially between the hips and feet. Through carefully constructed and meticulously executed interferences that balance out sinuous flow and decentralized pliability with abrupt juxtapositions and movements of contrariety, the ginga induces a disciplined relaxation, a rehearsed vertigo, a fluid and cool way of being-in-the-world. The endless juggling across these Africanist aesthetic principles outlined above informs the recuperation-cum-invention of renewed socio-historical identities and, in the last instance, promotes the articulation of inter-subjectivities in/as motion. Ginga has functioned, in that sense, as a non-hegemonic apparatus of enunciation: a critical reference for new modes of personal and collective identifications outside colonial thought.

Middleware

Having established the aesthetic elements associated with the ginga aesthetic, it is fair to assume that most people who grew up in Brazil have, to a certain extent, been exposed to and familiarized with these bodily dynamics. Am I saying that all Brazilians know how to swing their hips as professional samba dancers do? No, but ginga is a very popular and widespread concept within Brazilian culture. At the same time, is it possible to learn to discipline one's body to swagger like Brazilian footballers do? Yes, absolutely. Yet, before I unpack how one may step into this swayed action to articulate coolness with a play of hips, feet, and heels, as Rui Castro's quote opening this chapter suggests, I would like to take a moment to return to the question of why the standard middleware of the Brazilian Terrestrial Digital TV System is called Ginga.[11]

Ginga middleware (see the Introduction) is a high-tech product developed in Brazil (PUC-Rio and UFPB)[12] and implemented in partnership with Japan. The term "middleware" simply refers to a particular kind of software, sometimes called the "software glue," which lies between an operating system and multiple applications, e.g. interactive media. If you were to type the word "ginga" into your smartphone and perform an Internet search, for instance, Web crawlers would immediately scramble cyberspace, searching for answers, within an ocean of information "out there." Meanwhile, the search engine would also look up the personal data stored in your mobile "right here," gathering information about the habits and preferences associated with that machine's users (current location, language, previous searches, purchases, etc.). In doing so, middleware contributes towards mediating and personalizing the interaction between humans and machines, thus providing a smooth and continuous flow of information in both directions. As middleware for interactive TV, Ginga's job is to ensure that applications coming from a variety of sources work together seamlessly, constructing an environment that facilitates input/output interactivity between clients and servers. These non-linear synchronizations take into consideration predictable events (e.g. the duration of an episode) and unpredictable events (e.g. users' individualized interventions). Thus, ginga middleware promises a pliable way to acquire, process, and deliver data, while interacting with expected and unexpected phenomena, or spatio-temporal event-driven synchronizations, without disrupting the overall flow of the task at hand.

According to the developers' website, ginga is the fundamental movement in capoeira.[13] In terms of comparison, the choice of name seems to draw attention to the importance of investing in locally developed solutions (capoeira playing or software developing) in order to increase the nation's competitive impact in the global market. As an open source middleware for digital TV applications, Ginga middleware acts on two fronts: it provides good support for social/digital inclusion and interactive dissemination and acquisition of information. Hence, while capoeira's ginga is a core element of a marginalized practice of cultural resistance that spreads across the world from the ground up, the Brazilian middleware is a locally developed high-tech product that enables the inclusion of underprivileged segments of society and the open flow of information.

Of course, this is not the first time that this "Brazilian" swagger has been evoked to sell high-tech products and ideas. In the last few decades, capoeira's languid and off-beat way of moving, especially its acrobatic maneuvers, has been used in advertisement campaigns for products that offer versatility and innovation, such as fast cars, digital watches, and energy drinks. Yet, in the case of the Ginga middleware, there seems to be a deeper connection that gestures towards not only what ginga evokes in the consumer's mind (effects), but also the way in which it operates (functions) and the parts it plays (roles). While the ginga in capoeira structures how players use their bodies to improvise call-and-response interactions with other players (i.e. moving through seen and unforeseen events), the selling point of the Ginga middleware is its decentralized and multi-task-oriented architecture that is able to constantly adapt and rearrange itself in relation to the flow of information, while delivering a seamless performance or outcome. It manages how information flows across computers and interactive TVs (mechanical apparatuses). Hence, both the unit of movement and the middleware perform similar jobs: they organize the flow of information within a particular sphere, such as the bodies of capoeira players or the infrastructure of interactive media. In other words, one could say that the decentralization, flexibility, reversibility, improvisation, and continuous flow that the middleware's architecture offers its clients and servers working within the medium of digital TVs in Brazil finds resonance with (or mimics) the way in which capoeira players structure their bodies to dialogue with their partners, swaying their upper torso and hips to evade a blow or transforming an escape into an indirect attack. The slogan written in their logo says it all: "interactive TV is done with ginga."

This analogy makes sense. In the lineage of capoeira known as *capoeira angola*, for example, the multi-metered pace with which players rock their bodies back and forth, and the multi-linear pathways they construct as they tumble across the floor, alternating the use of feet, hands, and head to support their bodily weight, prepare them to shift directions and intentions seamlessly and with a sense of coolness. This decentralized and multi-metered way of negotiating space inside the circle makes it possible to "play apart," yet stay synchronized with multiple external elements (e.g. partner, music, and audience); to maintain an alert relaxation, while avoiding disruptive interactions, i.e. to attack and defend without losing one's corporeal sense of coolness. From this point of view, ginga may be better understood as the key element within an embodied program or system for interacting and exchanging ideas corporeally (i.e. the ginga aesthetic), anchored in concepts such as decentralization, flexibility, reversibility, and interactive improvisation.

Despite the similarities between the underlying logic structuring the Japanese-Brazilian middleware and the Afro-Brazilian ginga aesthetic orienting capoeira's call-and-response improvisations, their effects are not equivalent. Mechanical forms of apparatus such as smartphones and TVs process, disseminate, and store data but maintain its physical structure and its modus operandi intact. Conversely, the continuous articulation of the ginga aesthetic, or any other movement system for that matter, re-structures how players organize their moving bodies to think and act (*corpo-media*; see the Introduction).

In *Fenomenologia do Brasileiro* (1998),[14] for instance, Flusser offers a phenomenological observation about the African syncope in Brazilian culture, which enables us to better understand the extent to which the ginga aesthetic has permeated Brazilian society. Based on his personal experience living in Brazil from the 1940s to the 1970s, Flusser proposes that people born and/or raised in that country have, unlike migrating foreigners such as himself, a distinguished familiarity with the African concept of polyrhythms or multi-meter. Like Wisnik, Flusser's argument does not imply a genetically inherited aptitude, gift, or inclination to play traditional music from West Africa and accurately reproduce its rhythmic patterns. Rather, Flusser points to a highly specific structuring organization of bodily movement with which many Brazilians articulate syncopation as they move around in their day-to-day life. Flusser calls it a "structured gesture," which expresses dynamic beauty and sensuality through everyday life actions. When addressing the relationship between syncopated sounds and movements, he poetically infers that:

> In fact the fundamental rhythm does not manifest itself primarily in acrobatics, nor necessarily in "works of art" (which, like sambas and playful fights, are nothing more than epiphenomena), but in day-to-day gestures ... that radically distinguishes the Brazilian environment from others. The rhythmic gait of girls and young women, the dance-like footsteps of the boys on the street (accompanied by internalized look and smile, as if to manifest the power of rhythm over the spirit), the constant tapping on matchboxes and spoons, the use of typewriters in the offices as if they were drums machines, the transformation of hammers into *atabaque* drums, the grace of kids' gestures as they play football, and even the elegance of movements in street fights, all that is a manifestation of a profound culture ... The African syncopation and the highly (sophisticated) organization of bodily motion indicate that this is culture in a deep-seated sense, and this makes living in Brazil into a continuously experiencing of

> culture, though the fact is not always present. (Flusser, 1998, pp. 134–5, author's translation)

As described above, for Flusser, the influence of the African aesthetic in parts of Brazil is not limited to cultural and/or artistic products or "objects." Over the years, he proposes, locals have fostered artistic and cultural processes with and through the embodiment of aesthetic concepts such as the African syncope. Connecting motor and cognitive processes, for Flusser, such 'structured' behavior, cultivated in practices of everyday life such as typing, walking, and playing, has been absorbed within mainstream culture, thus transforming how Brazilians[15] approach life and behave in society.[16]

Paola Berenstein-Jacques' *Estética da Favela* (2001b) offers a detailed analysis of how the African syncope and its corporeal articulation, i.e. ginga, have distinctively informed how members of a particular community think and act. Berenstein-Jacques proposes that the work of the counterculture artist Hélio Oiticica (1937–80) departs from the aesthetic knowledges equally ingrained in the culture of the *favela* Mangueira, where Oiticica had invested a significant portion of his time and creative energy since the 1960s, and in the architectural structures of that community, made up largely of disenfranchised blacks and mulattoes. In doing so, she draws a relationship between the "aesthetics of ginga" (*estética da ginga*), which Mangueira's inhabitants cultivate through ephemeral actions, and the more permanent infrastructure of the favelas. She also relates the concept of ginga in samba to favelas' precarious make-shifts (*gambiarras*; see the Introduction), its dwellings, and its collective urban planning. Contrary to the prevailing belief that the architecture of Rio de Janeiro's favelas lack structural organization, Berenstein-Jacques suggests that these (illegally) claimed territories subscribe to a particular spatial identity and logic of organization. Insightfully, she proposes that the structural organization of favelas such as Mangueira both informs and is formed by the look and feel of ginga, as practiced in that particular locale:

> The experience of walking up or down a hillside shantytown [*favela*] is tinted by a unique spatial perception. As one passes by the first "breaking points" [*quebradas*], one begins to uncover a different walking rhythm, imposed by the alleys' pathway itself. This is what they call ginga. Wandering around the meanders of the favelas, we comprehend how those hillside children [*crianças do morro*] figure out how to dance samba way before they learn to properly walk. (Berenstein-Jacques, 2001b, p. 66, author's translation)

As I gather from the quote above, the architectural design of favelas such as Mangueira enables children to experience and incorporate, early on in their life, aesthetic principles associated with the concept of ginga (e.g. serpentine pathways) kinesthetically. It thus becomes ingrained in their muscle memory. At the same time, the embodied knowledge and the daily trajectories cultivated over the years by the inhabitants of the favela who are accustomed to think and act in such terms are archived in these community-built ethno-scapes. Closely connected to Flusser's observations of "Brazilians," Berenstein-Jacques' thick description makes a compelling argument towards understanding ginga as a common mechanism within a collection of inter-related products, processes, interactions, and ways of arranging "things."

Making a connection between bodily actions, social events, and practiced places, Berenstein-Jacques (2001b) argues that ginga is the common denominator of Oiticica's body-events, the architecture of Mangueira, and its music and dance (samba). The favela Mangueira, she argues, promotes a particular kind of spatial aesthetic of movement, i.e. movement-space trajectories. For the sake of clarification, Berenstein-Jacques organizes her analysis around three "conceptual figures:" the *fragment* (from body to architecture); the *labyrinth* (from architecture to urban planning); and the *rhizome* (from urban planning to territory).[17] Although she does not offer an in-depth analysis of ginga in dance performances typically associated with Mangueira, such as its famous samba school parades, one is able to better infer from her analysis of Oiticica's body of work how the underlying logic informing the cultural productions of that favela, from dance to architecture, are shaped by the aesthetic principles with which its inhabitants arrange their own bodies to move around and think about the world.

From where I stand, the favela territory may be understood as (in) formative sites – somewhere between a living and breathing archive and a choreographic score for collaborative and multiple improvisations. For those raised in the favela's "practiced place" (Certeau, 1984, p. 117), the physical interaction with the aesthetic principles structuring the "movement-space trajectories" of Mangueira has had a decisive influence in shaping how they organize their pathways. People incorporate the aesthetic knowledge structuring the space they call home into their own cultural processes and productions, from walking and dancing to music and architecture. Finally, the aesthetic figures that Berenstein-Jacques (2001b) identifies in both Oiticica's performative pieces and Mangueira's ethno-cultural organization of space reverberate this way of organizing things – i.e. the ginga aesthetic – with which individuals in that community work, play, make art, make shelters, or make do. As individuals who embody the ginga aesthetic recycle material, energy, and ideas to build and conquer new frontiers, this movement system is disseminated towards both micro (body) and macro (territory) directions.

Contrary to Hélio Oiticica's body-events and installations, the contemporary choreographies of Lia Rodrigues Companía de Danças, currently based in Rio de Janeiro's Favela da Maré,[18] offer little reference to samba dancing or other forms "typically" associated with Rio's shantytowns. Yet, like Oiticica's artistic collaborations with the inhabitants of the Mangueira neighborhood in the late 1960s, the underlying structure of Rodrigues' *Pororoca* (2009) and *Piracema* (2011), productions created in response to the choreographer's interaction with inhabitants of Maré, seems to resonate with Berenstein-Jacques' observations on the aesthetics of the favela in Brazil. In fact, when asked to comment on *Pororoca* during the Festival Panorama of 2010, Lia Rodrigues stated that:

> Pororoca bears in itself a way of moving and being that I learned, and admire, and like, that is the way of being in the Maré neighborhood. It is a different way of moving; it is a different way of shifting; it is a different way of inventing things, it is a different way of being-in-the-world, a rich and inventive way; a creative and inventive way. (Author's translation)[19]

At first sight, Rodrigues' recent choreographies may look like a mimetic reproduction of the favela's chaos, but this could not be further from the truth. Upon a closer look, one comes to realize that both *Pororoca* and *Piracema* reaffirm, and in some cases dialogue with, the fragmented, labyrinthian, and rhizome-like system of organization with which the residents of that favela organize their bodies and generate choreographed discourses. Rather than enacting movements of bodily syncopation on stage, I propose that Rodrigues' choreographies dialogue with Maré's underlying system of organization and production.

Defining the ginga aesthetic as a system of organization and production

Thus far, ginga could be better understood as the key element within a system of organization or logic, which exercises a distinct way of arranging one's body and interacting with one's surroundings. Contrary to the geometrical precision, radial linearity, coordinated unity, and upward mobility cultivated in various Western practices (i.e. classical ballet, gymnastics, and the waltz), the ginga aesthetic pushes individuals to imagine and reorganize their moving bodies as a *polycentric* structure in space, capable of articulating *polyrhythmic* patterns of movement in time. This underlying system of organization and production grounded in polycentrism and polyrhythm thus cultivates a pliable and articulated use of bodily parts, and a decentralized and multi-meter approach to the transfer of weight in time and space. As a dynamic modus operandi, the ginga aesthetic accommodates both abrupt ruptures and the continuous transition from one position to another, while avoiding fixed stances or static propositions. For instance, in the particular case of capoeira angola, practitioners look for ways to fragment their polycentric and polyrhythmic bodies into interdependent sections with which to unfold sinuous gestures, swift sweeps, fake strikes, and counter-attacking escapes. Similarly, in samba dancing, the ginga aesthetic enables dancers to articulate high-affect juxtapositions of the torso instantiated by a call-and-response dialogue between shuffling of the feet and gyration of the hips.

As discussed below, Figure 1.1 illustrates the primary elements associated with the system of bodily organization and knowledge production that I call the ginga aesthetic: (1) the propellers (feet); (2) the pairs of scales (hips and shoulders); and (3) the flexible vertical axis (the spinal column).

In order to cultivate the ginga aesthetic, one must (re-)structure one's body as a *polycentric unit,* aligned three-dimensionally around two major pairs of joints, namely the hips and the shoulders. Working like balances or *pairs of scales,* these two sets of horizontal intersections connect the polycentric core of dancers to their articulated limbs (arms/hands and legs/feet). In addition, these two pliable sets of joints are responsible for constantly rearranging and counterbalancing the weight of their body as they move in three-dimensional space. Moreover, one should imagine the spine as a *flexible vertical axis,* which functions as a dynamic line of communication between the two (horizontal) pairs of scales. The vertebral column aligns both the hips and shoulders perpendicularly and, subsequently, maintains a direct but pliable connection between the head and the tailbone (the sacrum and the coccyx).

The dynamic of this polycentric system produces specific and crucial effects. As indicated above, it dismantles the unity of the corporeal core, enabling its parts to move (i.e. shift weight) independently, while maintaining a constant flow between them. The articulated flexibility between the vertical axis and the two *pairs of scales,* sometimes described as a "serpentine" or "coil body" (*corpo de mola*), expands the subdivision

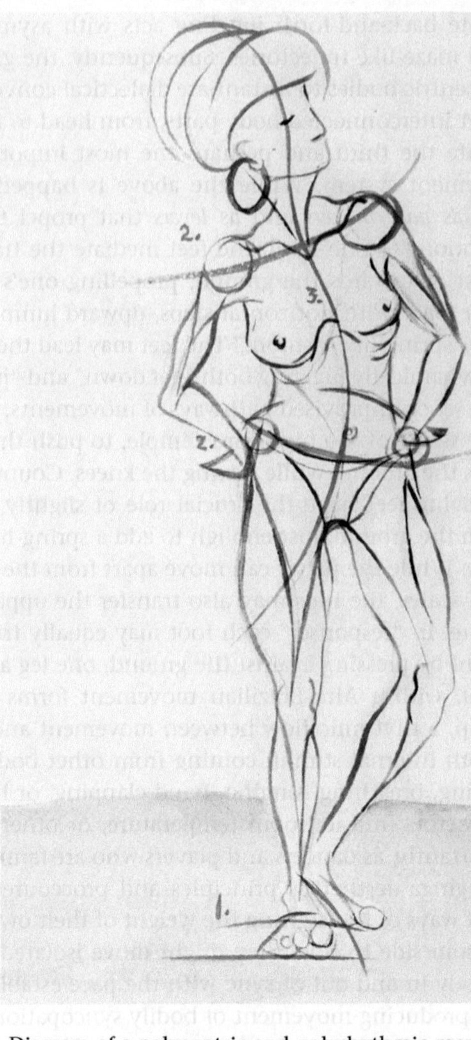

Figure 1.1 Diagram of a polycentric and polyrhythmic moving body © Cristina Rosa, 2014

of the torso into multiple centers. Hence, its gravitational center slides, back and forth, across different locations. In this dynamic process, the dialogue between the push-and-pull of muscles and the rotation of the joints expands the body's scope of possibilities for weight distribution, including asymmetrical configurations and curvilinear alignments. This flexible and decentralized alignment of the body also makes it possible to articulate back-and-forth juggling acts with asymmetrical weight transfer and maze-like trajectories. Subsequently, the ginga aesthetic enables polycentric bodies to instantiate dialectical conversations between isolated yet interconnected body parts, from head to toe.

The feet constitute the third and perhaps the most important element of this movement system. While the above is happening, the feet function both as *pace makers* and as *levers* that propel the body rhythmically in motion. At one level, the feet mediate the transfer of body weight against or towards the ground, propelling one's body in (three-dimensional) space with horizontal steps, upward jumps, downwards slides, or in a "spring-like fashion." The feet may lead the body to bounce vertically, dynamically juggling both "get down" and "jump up" modes of action, in set or improvised pathways of movements. A samba dancer may use the swing of the hips, for example, to push the weight of her body towards the ground, while flexing the knees. Counteracting such effort, the moving feet have the crucial role of slightly pushing the body away from the ground, just enough to add a spring-like riff to their rhythmic flow. While the pelvis can move apart from the torso, as the primary pair of scales, the hips may also transfer the upper body's weight as they swing. In "response," each foot may equally trigger the hips to slide outward by pressing against the ground, one leg at a time.

On another level, within Afro-Brazilian movement forms the feet dictate, and keep up, a rhythmic flow between movement and sound, dialoguing with both internal stimuli coming from other bodily parts (e.g. the heart beating, breathing, singing, hand clapping, or hip shaking) and external factors (music, room temperature, or other moving bodies). More importantly, as dancers and players who are familiar with or exposed to the ginga aesthetic's principles and procedures experiment with different ways of transferring the weight of their own bodies up and down, or from side to side, they might move isolated parts of their polycentric body in and out of sync with the pace established by the footwork, thus producing movement of bodily syncopation.

When I join a samba circle, for instance, my first concern is to listen, and feel, the pace of the percussive orchestra with my feet. With my feet hip-width apart, I typically transfer the weight of my body from one side to another with small steps in place. Once my feet are rhythmically connected to the basic samba beat (2/4), I then establish a line of communication between my feet and other isolated parts of my body from the ground up. As I keep up with the metric of samba, gently shuffling or dragging my feet back and forth, I may allow my hip joint to sway like a seesaw, rocking the waistline sideways, in and out of sync with the footwork (main riff). Rising up, this foundational dialogue between my feet and hips may also trigger the tailbone to oscillate, (mis)leading the flexible vertical axis into dynamic alignments, such as serpentine pathways. Similarly, when I enter a capoeira circle to play, I first connect my steps to the beat of the live orchestra. Yet, instead of shuffling my feet close together, as in samba, I maintain a wider stance as I construct triangular pathways across the floor (see Chapter 4). While this is happening, my hips and shoulders may work in tandem with the flexible spine to counterpoise or rearrange my body into transient balanced compositions, thus evading static geometric forms.

Furthermore, in samba, my hips have the primary function of swaying apart from my upper torso and feet, although they work in tandem with one or both of them. As my feet shuffle back and forth, for instance, my hips syncopate in and out of the 2/4 metric with flourishing wiggles and high-affect juxtaposition thrusts, thus drawing the attention (of participant-observers and the outside audience alike) towards the navel area (*semba* in Bantu). Due to the centralized position of the pelvis and its intra-connected range of motion, in most Afro-Brazilian dances the hips act as the primary pair of scales. While the hips offer structuring support to the upper body, the centralized position of the tailbone (the sacrum and the coccyx) in relation to the pelvic area connects the base of the flexible spine with this articulated center. At the same time, sitting above the upper thighbone, the hip joint connects the navel to the lower limbs towards the ground. Hence, as the feet propel the moving body back and forth, this primary pair of scales (hips) constantly rotates and re-structures the body's dynamics of balance. Within a certain lexicon of movements practiced in these forms, the swing of the hips may trigger the upper body to adjust its weight, shifting the flexible vertical axis (spine) and the secondary pair of scales (shoulder) into pliable twists and thrusts. At other times, the gyrating movement of the hips may equally interact with the legs, knees and feet, increasing their range of locomotion. Following this trajectory, once dancers familiarize themselves with how to arrange their moving bodies as polycentric and polyrhythmic structures, they are able to learn the particular lexicon of movements cultivated in practices centered on the ginga aesthetic, including, but not limited to, samba and capoeira (see Part II).

Embodying an idea or how can one get ginga

In the bossa nova classic "Garota de Ipanema" (1962), the music composer Tom Jobim and the poet and lyricist Vinícius de Moraes offer a more sophisticated way of drawing attention to the presence of the ginga aesthetic in everyday life as well as some of its most widespread metaphors. Perhaps the (English-speaking) reader of this book is familiar with the North American version of this iconic melody, "The Girl from Ipanema" (lyrics by Norman Gimbel), first recorded by Astrud Gilberto (1964), João Gilberto's wife, and Frank Sinatra (1967). (If you are not familiar with this, take a minute to look it up online and listen to it before going any further.)

In its original Portuguese version, Moraes inundates the listener's ears and mind's eye with a seaside scene that may be translated as follows:

> Look at this most beautiful and most gracious thing / She is the girl, who comes and passes by / With a sweet swing, on her way to the ocean.
> [*Olha que coisa mais linda, mais cheia de graça / É ela a menina que vem e que passa / Com um doce balanço a caminho do mar.*] (Author's translation)

Here, Moraes' verbal imagination slides into the pace of Jobim's offbeat composition. Or, rather, the Portuguese lyrics vivify the tonal syncopation – swing or ginga – first woven in Jobim's melodic abstraction, thus configuring a particular shape, color, and set of motions to its central leitmotif. Moreover, the muse's brief arrival and departure, the ocean where she is heading to, and her memorable "sweet swing" (*doce balanço*) all allude to the libidinous oscillation insinuated in the musical score. In the second verse of the Portuguese lyrics, the poet goes one step further and confesses that the girl's swayed action (*balançado*) "is the most beautiful thing that I have ever seen pass by" (*é a coisa mais linda que eu já vi passar*).

Now, to further illustrate my point, I would like to invite you, my reader, to try a practical exercise. Simply close your eyes for a minute and recall this bossa-nova tune in your mind. Hum it if that helps you to get in tune with it. Then, as you are, try to recreate the cadence of this melody by moving any part of your body, such as tapping your feet, snapping your fingers, or gently rocking your head from side to side. Get up and walk, if you can, embodying this infatuating rhythm as you gently glide your bodily weight from one foot to another.

When one follows this song up close, as I hope the reader was able to experience kinesthetically, one gets a sense of how "The Girl from Ipanema" invites the listener to move "like a samba that swings so cool and sways so gentle," as the English version suggests. In fact, the syncopated cadence of Jobim's side-to-side leitmotif (a mixture of samba, jazz, and other musical traditions) acts as a cartography over which the swaying hips of Moraes' idealized muse glides through with a sensual ebb and flow. Since both sound and movement resonate with the coming and going of Ipanema's waves and tides,[20] together Jobim and Moraes construct a triangular network of metaphoric relations, where syncopation is considered at the same time to be beautiful (the girl's swayed-walk), ephemeral (passing by), and sublime (Ipanema Beach). Clearly, in this context, the girl's body is imagined as an ideal – and thus unattainable – object of sexual desire. Borrowing Blanco Borelli's understanding of *hip(g)nosis* (2008), it nevertheless becomes evident that the song and the lyrics (re-)create a slippery dislocation of meaning across these overlapping mental imageries, all of which frame the swayed-walk of Ipanema's "it" girl – her "hip-notic" sweet ginga – as an auspicious local sight. "Ah if she only knew," Moraes concludes, "that when she passes by / the whole world is filled with grace / because of love" [*Ah se ela soubesse / que quando ela passa / o mundo inteirinho se enche de graça / por causa do amor*]. In the midst of the sexual revolution of the 1960s and the popularization of the two-piece swimsuit, this eloquent and sophisticated way of articulating ideas corporeally, as described in "Garota de Ipanema," is embraced as a source of local admiration and delight. Simply put, Jobim and Moraes repackage ginga as a beautiful bodily disposition mastered by girls from Rio de Janeiro (*cariocas*). Yet, as the next chapter will seek to elucidate, mainstream discourses where ginga evokes beauty and grace are a rather recent phenomenon.

2
Historicizing Ginga

> What's in a name?
> That which we call a rose
> By any other name would smell as sweet.
>
> Shakespeare, *Romeo and Juliet*

> It is not as much a question of what a word or a text "means" ... but of analyzing the way a word or a book functions in connection with other things, what it makes possible, the surfaces, networks, and circuits around which it flows, the affects and passions that it mobilizes and through which it mobilizes.
>
> Nikolas Rose, *Powers of Freedom*

I would like to take a moment here to recap my last few steps. In the previous chapter, I offered an introductory analysis of the embodied concept of ginga, connecting this sinuous and offbeat way of moving to the African presence in Brazilian culture and society. While I argued that anyone may, theoretically, step into the "Brazilian" swagger, I also postulated that the continuous deployment of the ginga aesthetic through movement shapes how people may be perceived and thus identified. Moreover, it also rearranges how they use their body to have fun, to interact with people, or to solve a problem. Over time, I proposed, the acquisition and continuous deployment of an embodied system of organization and production, including the ginga aesthetic, becomes instrumental to the ways in which bodies are wired to carry themselves, interact with others, or survey their surroundings. In other words, the (multiple) way(s) we choose or are coerced to both conceptualize and utilize our moving bodies, our individual and collective corporealities, inform and are formed by how we make sense of the world around us, how we walk around, and what happens when we move that way.

With this information at hand, in this chapter I investigate the complex negotiations between the word "ginga" and the corporeal metaphors insinuated through this swayed way of moving. I historicize in particular the overlapping concepts entangling moving ideas and their fixed significations in the Portuguese language. This chapter thus functions as a prelude of the primary issues surrounding both the literal and metaphorical meanings commonly associated with ginga. I should anticipate, nevertheless, that I simply cannot say all that can be said about this subject. What I offer below is a range of disembodied translations and re-significations regarding the noun "ginga" and its intersection with an embodied system of organization and production, grounded on aesthetic principles that have been circumnavigating the black Atlantic since the days of the international slave trade. By unpacking a number of conflicting ideas historically attached to this term across the Lusophonic (Portuguese-speaking) world, this chapter is designed to give the twenty-first-century English-language reader a taste of what is at stake whenever one hears, reads, imagines, writes down, or speaks the word "ginga," or makes reference to one of its multiple synonyms and dysphemisms. Finally, I should disclose that although this chapter begins with an etymological discussion of this term, the next few pages clarify for the reader not only the roots and anchoring retentions of this concept in the New World, but also – and most significantly – its transient routes, connections, ruptures and exchanges. To reference Nikolas Rose's quote above, in this chapter I investigate what the signifier ginga "makes possible, the surfaces, networks, and circuits around which it flows, the affects and passions that it mobilizes and through which it mobilizes" (Rose, 1999, pp. 29–30).

Defining ginga

In Continental Portuguese dictionaries, the noun "ginga" is first listed as a scull, an oar placed over the stern of a boat. It thus identifies an instrument or apparatus with which the oarsman (*gingador*) propels a boat with a side-to-side motion. The understanding of ginga as the act of tilting the body from side to side while walking or an enactment of bodily oscillation is listed second, most likely derived from the previous definition. The syncopated ebb-and-flow of the body commonly known as ginga (or gingado) in capoeira resonates with, or perhaps alludes to, the kinesthetic sensation of sculling a small canoe or stand-up paddling.

Like oarsmen, I propose, capoeira players tend to shift their bodily weight back and forth with a cool yet swaggering cadence that balances out strength and flexibility. In addition, the languid and offbeat momentum of the players recalls, as many capoeira songs suggest, the swaying motion of the sea, the riff of waves and the dynamic balance of tides.

Portuguese dictionaries also list a third meaning for the noun "ginga," which is connected to sugar cane plantations and is rarely used these days. In that particular context, ginga denotes a long-handled stick attached to a cup-shaped bowl used to transfer sugar cane syrup from one container to another. Similar to the first definition, here ginga means an instrument used to maneuver yet another liquid substance. Curiously, the oar and the long sugar cane ladle are connected respectively to two (out of three) macro structures that propelled Portugal into amassing a transatlantic empire during the sixteenth century: its mastery of the seas and sugar cane plantations in the New World. But that is not all. The slave trade system, Portugal's third and most powerful trump card, also seems to have indirectly played a role in the adoption of the term "ginga" to describe apparatuses handled with repetitive, circular and swing-like actions, such as rowing or moving syrup from one container to another.

In most Portuguese dictionaries the etymology of the word "ginga" is listed as "unknown." However, recent studies show that the term "ginga" is most likely derived from one of the Bantu languages. These correlated languages were historically spoken in the Kingdom of Kongo, where the Portuguese arrived in the early fifteenth century. Nei Lopes indicates in the *Novo dicionário banto do Brasil* (2003), for instance, that the verb *gingar* (to swing) is probably related to the Kimbundo (one of the Bantu languages) verb *junga,* meaning to spin from one side to another; the Kimbundo verb *jingala,* meaning to wave or to oscillate, from the same root of *jinga,* meaning to circle around, to stir up, or to spin around; or the Umbundo verb *yenga,* meaning to oscillate (Lopes, 2003a, p. 109). Julio Tavares also suggests that ginga derives from the concept of *jinga* in Mbundo, which is related to the idea of continuous motion and endless articulation (Tavares, 2002). He summarizes that across the Portuguese Empire: "Ginga became a figure of speech that indicates balance through movement, stylization, negotiation, non-confrontation and something like 'being in peace with God and life,' that is, a cosmic equilibrium" (Tavares, 2002).

Going back to Continental Portuguese dictionaries, one also finds that on the streets of nineteenth-century Lisbon, the noun *gingador* was informally used as a synonym of the noun *fadista,* then a derogatory term that referenced individuals of low class or questionable reputation (e.g. prostitutes, rioters, and thugs), who sang and danced to the Afro-Portuguese Fado.[1] Meanwhile in Rio de Janeiro, the noun *fadista* is also extended into the nineteenth century to figuratively describe male rogues and gang members, largely non-white men, some of whom were involved with *capoeiragem*, as capoeira activities were referred to at that time. Like thugs and fadistas in the streets of Lisbon, the capoeira players in nineteenth-century Rio de Janeiro, or simply "the capoeiras," were known to brandish razor blades as their preferred weapon and to perform a swayed gait as their signature way of moving. By word of mouth, this colloquial approximation or equivalence amongst oarsmen, bohemians, hustlers, and courtesans across the Portuguese Empire forges a metaphoric entanglement between movements of bodily swing or oscillation and the vacillating behavior of persons of "sinister," "dark," or "deviating" behavior.

Queen Ginga

The word "ginga" has carved another parallel route across the Lusophonic world. Within the context of the Portuguese colonization of Africa, the capitalized word "Ginga" is frequently used in reference to Njinga Nbandi Ngola (1582–1663), the legendary Bakongo Queen of the Kingdom of Matamba, today a part of Angola.[2] According to legend, Njinga was born with the umbilical cord twisted around her neck (a popular sign of defiance), hence her name.

As Selma Pantoja (2000) explains, the seventeenth-century African Queen was an unusually powerful and controversial persona, who led a historic resistance against the early Portuguese (and Dutch) invaders of Central Africa. For the historian, Ginga escaped a conspiracy before she could inherit her father's crown in the Kingdom of Ndongo. She then traveled east to the Matamba region (also known as the Jinga region), where she aligned herself with a warrior people known as the Jagas, adopting their worldview and eventually becoming their "iron-fist" queen. This larger-than-life figure fought against and made alliances with European colonizers until she was in her sixties. While she skillfully led troops in battle, she also traded local war captives and prisoners with different European expeditions. In 1622, she met with the Portuguese governor João Correia de Sousa and his entourage in Luanda, impressing them as a lucid and articulated political chief who could speak their idiom. Ginga converted to Catholicism – and was baptized as Ana de Sousa – as part of peace negotiations with the Portuguese.[3] Nevertheless, according to some accounts, she later resumed her "pagan" religious practices.

Despite the discrepancies surrounding the life of this unusual figure, it is safe to assume that Queen Ginga's unconventional diplomacy, especially her artful ability to offer resistance against political, economic, and religious means of subjugation, has influenced the ways in which her European contemporaries would described her. In 1669, for instance, the Italian Capuchin Frey Antonio Gaeta from Naples published a first-person account of Queen Ginga's conversion to Catholicism,[4] where the *Regina Singa* is characterized as the (until then) undefeated "enemy of Europeans." Alencastro points out that by the time that colonial accounts of that region were disseminated through the European printed press, the story of Queen Ginga is "transformed into the paradigm of barbarity in Africa, in the spectrum of chaos in which the overseas pagans lived. And the Capuchins brought them back to the Kingdom of God" (2000, p. 278). According to Gaeta, the "marvelous" life of this iron-fisted queen was filled with bizarre and sinful behaviors, including transexualism, cannibalism, and sorcery. Similarly, in Antonio de Oliverira de Cadornega's *História Geral das Guerras Angolanas* (*General History of Angolan Wars* (1942)), originally published in 1680–1, the "bizarre" Queen of Matamba is again described as a "masculine" figure who maintained a harem of young men forced to dress as women and act as her spouses.[5] Furthermore, in the Italian Capuchin missionary Giovanni Antonio Cavazzi da Montecuccolo's account of the Kingdoms of Kongo, Matamba, and Angola (1687),[6] Queen Ginga (spelled *Regina Zingha*) is depicted as a diabolic character with unimaginable vices that, despite her formal conversion to Catholicism, intertwined sexual perversion with physical cruelty.

In the following centuries, the "marvelous" life of Queen Ginga, and her "monstrous" black female-led kingdom in Africa, gained an independent life in subsequent European publications. In 1792, for instance, the Franco-Portuguese poet Manuel Maria Barbosa du Bocage published an insulting sonnet, "Lembrou-se no Brasil bruxa insolente," attacking the Brazilian poet Domingos Caldas Barbosa (Bocage, 1908, p. 94). There, Bocage characterizes Barbosa's mother, a freed slave from Angola, as a descendent of Queen Ginga, the "insolent witch" (*bruxa insolente*) who transforms primates into (insane and immoral) vassals. Queen Ginga also appears in the writings of the libertine the Marquis de Sade (1795) and the historian Georg Hegel (1822–3). While Sade cites Queen Ginga as an example of how Africans entwined savagery and eroticism, Hegel evokes her name to illustrate the primitivism (or delinquency) he attributed to Africans.[7] Grounded on what Quijano and Ennis calls a modern/colonial "idea of race" (2000), these examples above attest to the significant role that this "out-of-this-world" black female figure that devoured people, either literally or metaphorically, acquired in the construction of the African inferiority in relation to the European inter-subjectivity.

While these ideas circulate across Europe, in Brazil new tales of Queen Ginga re-emerge inside colonial pageant-parades sponsored by black-only Catholic Brotherhoods and Sisterhoods. Yet, before I consider the re-enactment of Queen Ginga's legend by captured, sold, enslaved, and converted Africans in the New World, let us first examine the context under which the noun "ginga" gained visibility in Brazil as a languid and offbeat swing connected to "dark," "primitive," or "sinister" bodies.

Ginga in Brazil

It might be impossible to elucidate how the noun "ginga" went from denoting apparatuses handled with an oscillating motion to connoting an embodied mechanism (or *dispositif* in the Foucauldian use of the term) that sparked off one's body to sway in a similar fashion.[8] What we know is that by the nineteenth century, newspaper writers began to employ the noun "ginga" sparsely to qualify a tottering, feeble, or unsteady kind of gait, often associated with the "inferior" bodies of enslaved Africans and their descendants. Movements of bodily undulation and/or oscillation were perceived, more widely, as distasteful, suspicious, or defective behavior, inasmuch as it digressed from the "normal," elegant, or civilized gait cultivated in Western Europe. In 1890, for example, the word "ginga" appears for the first time in the newspaper *Estado de São Paulo* in the description of a murderer that had a "slightly swayed gait" ("ginga um pouco no andar," 20 August, p. 2). Ginga reappears again in the same newspaper in 1915 in reference to a suspicious countryside-man tottering down the streets ("ginga pela rua," 16 February, p. 4). Over time, ginga comes to signify a swayed way of carrying oneself, commonly associated with a "dishonest" or "negative," and thus not trustworthy, character or essence.

Constructing shame

The lack of archival information connecting the noun "ginga" to this Africanist bodily disposition points to the fact that throughout the era of slavery in Brazil (1550–1888), colonial administrators, members of the clergy, and slave owners alike paid little attention, and perhaps felt a certain denial or repulsion, towards African life and culture in captivity. For instance, the Portuguese tended to label all activities that involved drumming and/or percussive rhythms as "batuques" (a pejorative term meaning noise or disturbance). At the same time, though the word "ginga" is absent from colonial archives, their descriptions of "batuques" featuring chanting, clapping, and dancing make reoccurring references to movements of "feet-and-hip agitation" (*saracoteado*), "bodily undulation" (*meneios de corpo*), and "serene oscillation" (*bambolear sereno*).[9]

As previously stated, the care of the self-exercised across African and Afro-Brazilian dances, rituals, and games clearly opposed the (normative) asceticism of the European/Christian culture of the self.[10] Whether practiced in black spaces such as slave headquarters (*senzalas*) and maroon communities (*quilombos*) or during public celebrations such as births, weddings, and funerals, in isolation or joined by other ethnocultural groups, these so-called "batuques" infringed upon the Western religious system of beliefs and bypassed or evaded their colonial codes of morality and civility. Movements that involved the "lascivious" gyration of the hips or the provocative "navel greeting" (*semba* or *umbigada* – see Chapter 3), in particular, exceeded or differed from the colonial-Christian sense of corporeality, including its expectations for bodily alignment and dislocation in time and space as well as its articulation of concepts such as beauty/grace, dignity, and decency.

Contrary to social activities organized under colonial-Christian morality or "honest entertainment" (*diversão honesta* – Dias, 2001, p. 859), the Africanist movements of bodily syncopation, sometimes described as "having the devil in the body" (*com o diabo no corpo*), were recognized as "dishonest" and thus shameful behavior.[11] In this regard, Monteiro (2002) states that it was not rare for priests to beg slave owners to restrain their enslaved Africans from dancing in front of saints' images inside Catholic churches, for their dancing style represented a threat to the redemption of their own souls and the souls of those witnessing these "immoral" and "profane" acts. Still, despite the generalized animosity and disdain towards "all things black" (*coisas de preto*), under the right circumstances, the populace was not only allowed but was also encouraged to dance as they pleased. During exceptional occasions such as religious and civic holidays, when the enslaved were granted a day off, "profane" music and dance was tolerated as escape-valve measures or "little evil" against bigger threats, such as a slave uprising (Reis, 2003).

European traveling artists who joined scientific expeditions across the New World during the Enlightenment era offer additional cues regarding the African/Africanist qualities of movement practiced in colonial Brazil.[12] Unlike the Catholic Church and the local authorities, these visual artists paid detailed attention to the ways in which these ethno-cultural bodies moved while dancing, fighting, or performing daily chores such as balancing water vases or food trays on their heads as they walked on the streets. Like colonial administrators and the priests, however, these European-trained craftsmen exerted a colonial gaze toward their subject matter, tinting their picturesque depictions with a veneer of exoticism/eroticism. Johann Baptist von Spix's *Die Baducca* (1817–20), for instance, illustrates an outdoor black dance accompanied by percussive live music with instruments of African origin. In this scene, dancers are represented with grotesque facial features and their joints bent like those of primates. In particular, Spix depicts black women with their hair untamed and their breasts exposed, suggestive of their "primitive" or "shameless" character.[13] In the lithograph *Batuque* (1835), however, Johann Moritz Rugendas presents a more stylized depiction of this lascivious colonial dance, with *beau savages* posing in the wild with supple and elegant *contrapposto*.[14] Nevertheless, in his traveling account, Rugendas explains that, besides the rhythmic clapping of hands, batuques "consist in certain bodily movements that might seem too expressive," from which he highlights the "agitation of the hips" (1954, p. 154).

In both examples given above, despite the artists' anatomical concern and their careful attention to detail, the epistemic racism embedded in these "naturalistic" artworks produces a depreciative imagination of the subjectivity of Africans and/or the aesthetic knowledges informing their ethno-cultural practices. Not unlike the fate of Queen Ginga's "marvelous" life, once these racist illustrations and traveling accounts about the "lascivious" black dances from Brazil spread across Europe, they invariably contributed to archiving (i.e. fixing) the ginga aesthetic as "proof" of the racial inferiority of its practitioners or the weakness of their character. Inside "cabinets of curiosities" (*Wunderkammern*), these artful representations of bodily "lasciviousness," from the "vicious" contortion of the torso to the "immoral" agitation of the hips and feet, come to *stand for* exotic/erotic otherness. By the nineteenth century, European dance experts such as the ballet master Carlo Blasis relied on these colonial accounts to exhort that the bodily contortion and agitation observed in dances from Africa and the Americas (i.e. the polycentric and polyrhythmic alignment of the core) not only deviated from the notions of beauty and taste he associated with the European/Italian "art of dancing" (i.e. ballet) but, along with the tropical climate of these regions, were also powerful enough to corrupt one's character (Blasis, 1976 [1831]).

Reconstructing pride

Nevertheless, from a decolonial perspective, the written documents and artworks mentioned in the previous section constitute an entry point from which to articulate the following premise: the nearly six million Africans trafficked into Brazil over the course of three centuries did not come "empty handed." Within their bodies, they bore particular aesthetic and philosophical knowledges with which to *re-create* the experience of their homeland as well as to *recuperate* their embodied sense of self-worth and belonging.[15] Moreover, although confined to the inhumane circumstances of subjugation and forced to perform or otherwise interact with different systems of organization, production, and belief, many drew upon their embodied memory to enact physical (and metaphysical) activities that could help them recuperate-cum-invent their sense of who they were as a people. Investing in a type of labor that produced both physical pleasure and ethnic pride, the enslaved and freed Africans' deployment of embodied knowledges connected to their ethno-cultural heritage in the New World contributed to restoring their sense of dignity and self-esteem.

In Part II of this book, I take a closer look at the centrality of the ginga aesthetic in two movement practices closely connected to Bantu heritage in Brazil: samba-de-roda and capoeira angola. Nevertheless, it is important to note that this movement system permeates nearly all Afro-Brazilian enactments of sociability that bear syncopated music and dance. Movements patterns commonly associated with ginga, such as movements of bodily undulation (*meneios de corpo*), hip-wiggling (*requebro*), hip-and-feet agitation (*saracoteado*), and the swing-walk historically known as "to sift" (*peneirar*), appear at a wide range of regional Afro-Brazilian rituals and dance practices, including *batuque, jongo, samba de umbigada, tambor de criola, jurema, maculelê* and *coco,* as well as other fight-games such as *samba duro* and *pernada.*

Movements of bodily syncopation or gingado have also infiltrated a range of colonial hybrid forms, such as pageant-parades sponsored by black-only Catholic Brotherhoods and Sisterhoods (*Irmandandes*) in Brazil. These colonial pageant-parades, also known as *folias*[16] (from the French *folie,* meaning madness), are particularly relevant inasmuch as they will function as the model for the development of modern samba school parades addressed in Chapters 3 and 5. Kongo King parades

(*Congadas*)[17] are one of the earliest and most widespread kinds of folia practiced in colonial Brazil. Despite their regional variations – *Congada, Folia de Reis, Rancho de Reis, Reisado, Cucumbis* or *Moçambique* – these dramatic dances aimed to illustrate the triumph of the Catholic missions and the conversion of African nations.[18] As Alencastro summarizes, Congadas juxtaposed the image of a (good) king of Kongo – positive, pacifist and Christian ally – against a (mischievous) Queen Ginga of Matamba – a negative, pagan, and invasive warrior (2000, p. 281; see also Câmara Cascudo, 2001).

According to the historian João José Reis (1996, 2003, 2005), these colonial black-only Irmandades were hermetic organizations, whose membership was based on gender, bio-ethnicity (i.e. language groups and religious beliefs), nationality (birthplace in Africa), or socio-political alliances in the diaspora. The conversion of blacks to these Catholic organizations indirectly contributed to the formation of coherent groups of individuals taken away from their original communities and families in Africa.[19] When these ethnic cohesive groups organized their folias, while the overall theme of their parades followed liturgical scripts imposed by the Catholic Church,[20] the rhythms and moves they enacted in between dramatic scenes often reflected the cultural specificities of particular African nations. Despite the imposed devotion and the controlling doctrinairism characteristic of these folias, which were reminiscent of Catholic celebrations during the Middle Ages in Europe, these spectacles of revelry called for community cohesion, which ensured the production of rich and lavish costumes, ostentatious banners, and after-parties. Furthermore, the family-like infrastructure of black-only Irmandades such as Our Lady of the Rosary of Blacks (*Nossa Senhora do Rosário dos Pretos*) provided, beyond the "salvation of their souls," a place, a time and an excuse for individuals of similar roots (e.g. language groups, birthplace, or religious beliefs) to interact and bond, re-creating a sense of community in the diaspora.

J.J. Reis (2005) affirms that during the slavery era in Brazil, these public performances of revelry, from "dishonest" batuques to "honest" proto-Catholic parades, functioned as "a basic reference of black and slave identity, so long as we remember that identity is not a fixed point in the experience of a group." He further concludes that:

> Identity can change and be multifarious. What is constant is a sentiment of alterity, collective singularity and often opposition. That is the reason why all black celebrations under slavery, albeit some more than others, represented a means of expressing slave and black

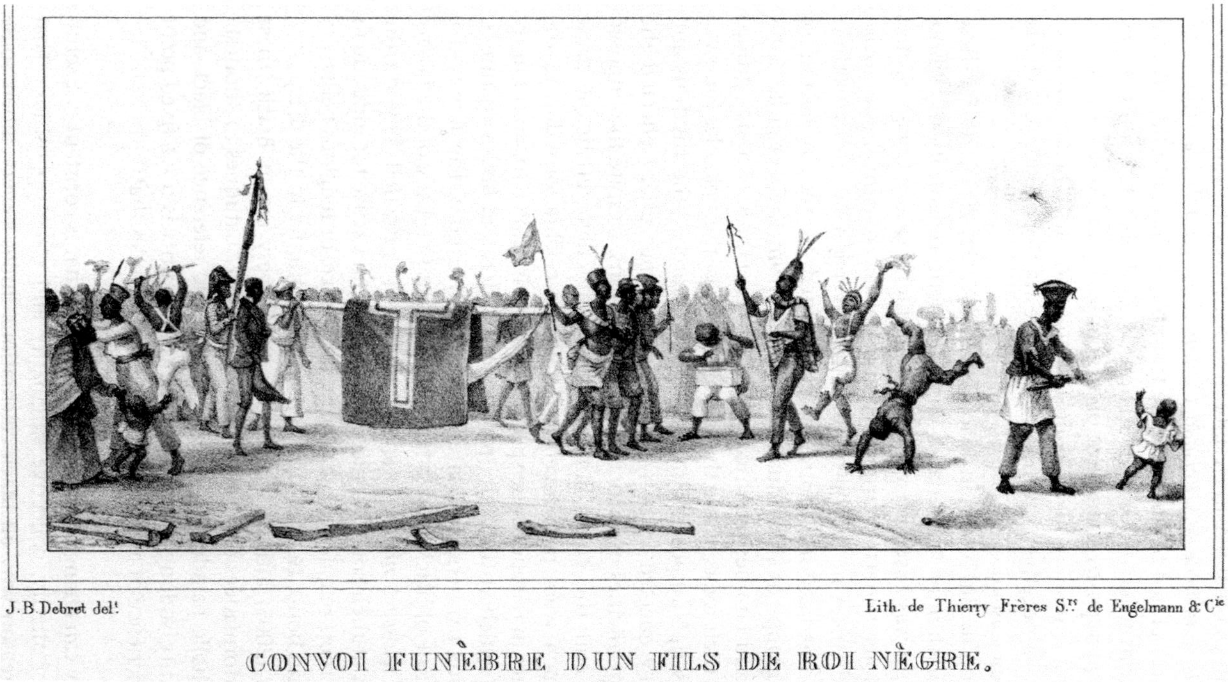

Figure 2.1 Jean Baptiste Debret's *Convoi funèbre d'un fils de roi nègre*, 1839

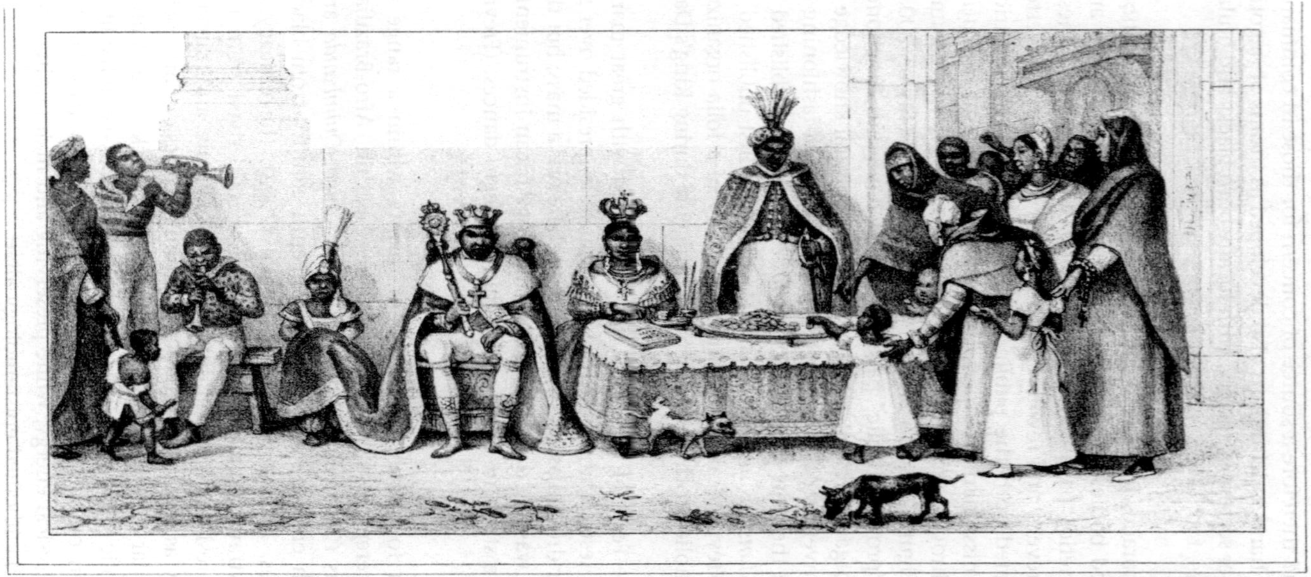

Figure 2.2 Jean Baptiste Debret's *Quête pour l'entretien de l'eglise du Rosario*, 1839

> resistance, and therefore a source of concern for those in command. On the other hand, celebrations also developed as means of negotiation with other sectors of society, with locally born blacks and mulattos and whites, too. (2005, p. 202)

These colonial parades further engendered transcultural juxtapositions between enslaved bodies and noble characters, European fashion and the African aesthetic, "pagan" excess and ecclesiastic epiphany, honesty and indecency, revelry and revolt. Over time, Reis concludes, converted blacks appropriated both the public (festive) space and the theatrical devices of the missionaries to perform honorific roles and *re-imagine* Africa in Brazil through choreographies, musical compositions, costumes and props (see Figures 2.1 and 2.2). As "acts of transfer" (Taylor, 2003), many of these "proto-Catholic" pageant-parades have thus functioned as rituals for re-organizing collective memory and cultural knowledge as well as for (re-)inventing their collective processes of self-fashioning or identification in the New World. In *Sobrados e Mucambos* (published in English as *The Mansions and the Shanties*), which was first published in 1936, Gilberto Freyre recalls a childhood memory of a folia organized by the Afro-Catholic Brotherhood of Saint Iphigenia during King's Day:

> There, on the Feast of the Magi, they celebrated with great marry making their festival, more African than Catholic, presided over by the old leader dressed as a king. To be sure there was a mass; but the main feature was the dancing to the sound of African instruments. Dances in the street in front of the church, Negro dances. (Freyre, 1963, p. 40)

Ginga is also a foundational movement quality within a range of incorporation ceremonies and praise dances within Afro-Brazilian religions such as *Candomblé, Tambor-de-Mina, Xangô, Quimbanda* and *Umbanda,* which emerged across the nineteenth century from black-only Irmandades. According to Rita Amaral (2002), the efficacy of Candomblé's extravagant celebrations called *xirês* (derived from the Yoruba ere/sere) resides in their ability to restore, cultivate, and transmit cultural memory as well as to produce communal cohesion and social networks.[21] Similar to what happened in colonial black social dances and proto-Catholic folias, these labor-intensive and body-centered liturgical practices encourage the lavish expenditure of resources and the dissipation of physical energy, while preserving and maintaining their (non-hegemonic) embodied epistemology alive.

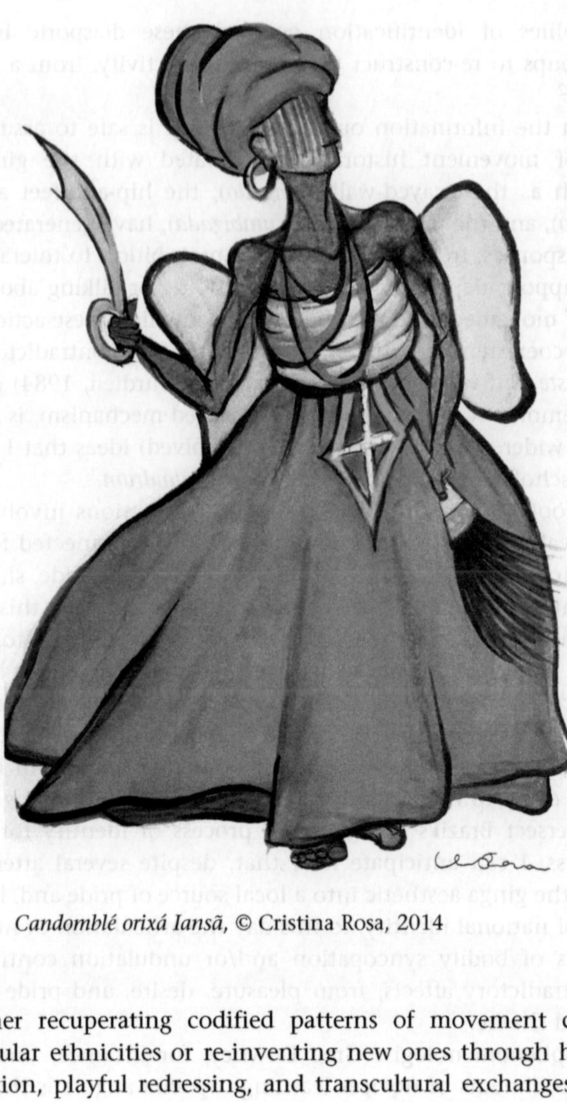

Figure 2.3 *Candomblé orixá Iansã,* © Cristina Rosa, 2014

Whether recuperating codified patterns of movement connected to particular ethnicities or re-inventing new ones through hyperbolic imagination, playful redressing, and transcultural exchanges, the distinct yet interconnected movement practices listed above have, since colonial times, empowered marginalized individuals with cultural agency and capital. More importantly, in all of these diverse scenarios, the ginga aesthetic has functioned as a central *apparatus of enunciation*, through which participants structure their bodies to produce (flexible) choreographies of identification. Combining movement patterns with musical rhythms, attires and hairdos, these set and improvised choreographies of identification enabled these diasporic individuals and groups to re-construct their inter-subjectivity, from a place of otherness.[22]

Based on the information outlined above, it is safe to assume that qualities of movement historically associated with the ginga aesthetic, such as the swayed-walk (*gingado*), the hip-and-feet agitation (*saracoteado*), and the navel greeting (*umbigada*), have generated a wide range of responses, from persecution and prohibition to tolerance and declared support, depending on who was doing or talking about these qualities of movement and the context under which these actions took place. The coexistence of these multiple and often contradictory *judgments of taste* and *value* (cultural capital; see Bourdieu, 1984) given to a non-hegemonic way of moving, or embodied mechanism, is symptomatic of a wider set of complex (and unresolved) ideas that I identify across my scholarship as a *pride-and-shame conundrum*.

As one looks deeper into the complex negotiations involving the perceived value allocated to a bodily mechanism connected to blackness (i.e. ginga), as well as the conflicting affects (i.e. pride, shame) its deployment has generated, one comes to understand that this conundrum stems from a deep-rooted paradox that is specific to Brazil's patriarchic colonization: the sexual and cultural interactions between races (miscegenation) inside a country socio-economically built on a racial division of labor (slavery). Below I take a closer look at the genealogy of ginga in capoeira since Brazil's independence from Portugal in 1822 in order to further illustrate how this way of moving continued to intersect Brazil's post-colonial process of identity formation. Nevertheless, I can anticipate here that, despite several attempts to transform the ginga aesthetic into a local source of pride and, later on, a symbol of national identity, to this day the articulation of Africanist movements of bodily syncopation and/or undulation continues to incite contradictory affects, from pleasure, desire, and pride to fear, anxiety and shame.

For the phenomenologist Greg Downey, for instance, one of the biggest impediments of novices learning capoeira angola is the shame connected to blackness. Although Downey does not deny that physical weakness and psychological fear may slow down beginners' learning process, based on his ethnographic study in Salvador, Bahia, he concludes that this "shame" causes the beginners' body to tense their muscles, for example, inhibiting their ability to "rediscover potential movement that our reservations led us to relinquish" (2005, p. 198). "Tearing out the shame" from student's bodies, he concludes:

Transforms the way they move and see life, no matter what their color; but it may be especially important to dark-skinned people in Brazil, who are told repeatedly by the media that they are ugly. By training Capoeira Angola, devotees strip shame from their bodies where dominant ideals of beauty suggests they should feel embarrassed; they become soft, supple, and sly, even though, outside the roda, they may be told that they are brutes, inelegant, and unthinking. (2005, pp. 198–9)

The pride-and-shame conundrum and the ginga in capoeira

The history of the practice of capoeira is entangled with heroic stories, starting with, as Matthias Assunção reminds us: "Black slaves fighting policemen on the squares of colonial cities in the New World, and gangs of 'tough guys' promoting mayhem, terrorizing citizens or helping corrupt politicians to rig elections" (2005, p. 2). Generally speaking, prior to Brazil's independence, "negro war dances" and "mock combat games" were recorded largely as primitive, occasionally violent (towards one another), and bizarre acts of revelry, through which enslaved and freed Africans attempted to solve matters "with their own hands," to exert influence over a particular geographic territory, or to simply play-fight between work shifts.

During Brazil's imperial era (1808–89) and specially in the decades leading up to the abolition of slavery (1888), the so-called "educated elites" in urban centers became increasingly hostile towards, and in some cases fearful of, the overwhelmed black presence within their cultural everyday life. This was especially true of economically striving urban centers such as Recife (a major port city) and Rio de Janeiro (the imperial capital), where the Portuguese court was transferred to in 1808. The escalating violence associated with, or attributed to, capoeira men organized in gangs (*maltas*)[23] in Rio de Janeiro, such as the *Nagoas* and the *Guaiamus*, led to the growing recognition of *capoeiragem* as a social/moral outbreak. In this context, the defiant "ruffian dancers" who defiantly swayed their bodies as they walked on the streets waving razors and daggers in their hands were collectively framed as deviant characters who disturbed civic life and threatened to harm others. Conversely, for the (largely non-white) capoeira players circulating in post-colonial Brazil, ginga or gingado continued to be employed as a non-hegemonic mechanism of resisting power-knowledge (in the Foucauldian use of the term)[24] with which to perform masculine honor and ethno-cultural pride.

Shortly after the abolition of slavery in Brazil, the battle to eliminate or at least contain this alarming social/moral "disease" reached a climax with the institution of the Penal Code of 1890, under which "exercises of corporeal dexterity and agility known by the name of capoeiragem" were recognized as a crime. As the Penal Code's text explains, anyone accused of running around causing fear or disorder, especially if caught with a razor blade or dagger in his possession, could be sentenced to two to six months in prison. In practical terms, the so-called "fear of capoeiras" became so widespread that scholars such as Soares (2004) suggest that the simple display of "gingado" on the streets could be considered sufficient proof for an arrest to be made. It is true that Rio de Janeiro's imperial police had a long history of arresting and punishing those engaged in capoeiragem, but with the abolition of slavery and the institution of the Penal Code of 1890, the authorities were finally able to detain and punish any razor blade-armed troublemaker without getting into conflict with slave owners. Under the scientific racism discourses in vogue at the time, which Schwarcz (1993) defines as "the spectacle of the races," the institutionalized violence towards non-white bodies (and their practices), in the form of persecutions, corporeal punishments, incarcerations, and exile, was prescribed as an aggressive, although necessary, *remedy* against that "open wound."[25]

The increasing persecution of capoeira/capoeiragem during the First Republic (1889–1930) was symptomatic of a larger issue assaulting Brazilian society at the time. Briefly, despite the abolition of slavery, non-whites – the vast majority of the population – continued to be treated as second-class citizens. In addition, the widespread miscegenation of Brazilian society had become a topic of national (and international) concern, for it was perceived as a hindrance to the development of a unifying "national spirit" in this former slave colony. In the end, the insoluble puzzle or dilemma curbing the European-aspiring elites' projection of a unifying sentiment in and about Brazil could be summarized as follows: how could they formulate a national sentiment in a post-colonial society built on the guilt of slavery and doomed by the shame of sexual/cultural miscegenation? How could they build a modern nation, following the Enlightenment's notions of progress and humanism, without questioning the exclusionary dualism of race relations cultivated in the former colony and, subsequently, their own privileged position? The inability of the elites to fully resolve these questions led them to a kind of post-colonial "identity crisis," which outlasted the First Republic.

From the other side of the spectrum, the insoluble puzzle haunting non-white men in post-slavery Brazil, whether they were capoeira players or not, may be summarized as follows: how could these individuals construct their inter-subjectivity within the public sphere, i.e. be perceived as subjects in civil society, when (the image of) their thinking-and-moving bodies and their ethno-cultural heritage continued to be discursively marked with otherness and socio-economically fixed within scenes of subjection?[26] From this point of view, the practice of capoeira in post-colonial Brazil functioned as a non-hegemonic structuring attitude or a way of life that, similar to modern North American practices such as hip hop, is connected to the self-esteem struggle of Africans and their descendants in the diaspora (Rose, 1996). Although marginalized, the resisting cultivation of capoeira after the abolition of slavery and the institution of the Penal Code provided disenfranchised young men with a protective and permissive arena to publicly display their physical strength, courage, virility, analytical dexterity, coolness, etc. Furthermore, the capoeira players' conscious articulation of aesthetic and philosophical knowledges connected to Afro-Brazilian heritage in their fight games and their praise of African and Afro-Brazilian warriors and visionaries from the past within their oral history[27] worked towards the decolonization of their bodies from archives and repertoires of subjugation (Abib, 2004). Subsequently, the practice of capoeira during the First Republic continued to offer these marginalized individuals a protective and permissive sphere where they could acquire the respect and admiration in their community, while reconstructing a sense of self-worth and dignity towards their intangible heritage. It becomes clear that, as Sansone proposes, capoeira has functioned as "another tradition which is very Brazilian and whose cultural text and actual practice offers an alternative mirror of black masculinity" (1999, p. 31).[28] In particular, the capoeiras' widespread adoption of the gingado-walk as their signature gait, i.e. an improvised, though carefully rehearsed, way of "being-in-the-world" (Heidegger, 1962) stood as a performative embodiment of their set and improvised processes of ethno-cultural identification in and out of capoeira circles. Ultimately, as will be further discussed in Chapter 4, the particular *way* players move, i.e. their signature style or choreography of identification, functions as an apparatus to negotiate power relations in addition to the way they look or their class status.

It is equally important to note that the increased surveillance and persecution of capoeiragem across the nineteenth century did not completely eradicate this Afro-Brazilian practice. Certainly, many capoeiras restricted their public activities to community gatherings located on the outskirts of urban centers. This further restriction of their field of action also contributed towards the docilization (and in some cases fetishization) of the practice of capoeira. Nevertheless, the often romanticized dandyism associated with the way in which these swaggering ruffians carried themselves publicly, or were imagined to maneuver through the city, also enticed eccentric members of the higher classes to discreetly, and sometimes overtly, learn and practice capoeiragem.[29] Within these elitist circles, the "forbidden" – and almost extinct – "negro war dance" is auto-exoticized as a mixed-race urban practice. At the turn of the twentieth century, for example, *the mulatto capoeira* resurfaces in newspaper chronicles, art magazines, and fictional novels, where this figure is depicted as "a fashionable rogue" or "a tropical dandy." Capoeiras were often stereotyped as vulgar or cunning, yet charming and dexterous, rascals (*malandros*).

In the early part of the twentieth century, an increasing number of avant-garde artists and intellectuals began to extend a voyeuristic interest towards "popular" or "folk" (i.e non-European or mixed) cultures. Following precursors such as the Afro-Brazilian historian Manuel Querino (1851–1923), whose earlier works (1909, 1914, 1916) sought to outline the contributions of Africans and mixed-race populations in the formation and development of Brazilian culture and society, the local intelligentsia attempted to formulate sporadic discourses aimed at endorsing the incorporation of cultural elements unique to Brazil within their collective imagination of the nation-state. In this new light, few physical education enthusiasts and military officers sought to repackage capoeiragem as a national regimen of training that could be re-directed to produce healthy citizens and servicemen. Amongst these initiatives, in 1907, a naval officer (an anonymous O.D.C.) published the first physical education guide to promote capoeira as a "national gymnastic."[30] In 1916, the Captain Ataliba Nogueira and Lieutenants Lapa and Leite published another manual of capoeira intended for military personnel only. Like many others that would follow, the underlying idea behind these handbooks was to codify a local modality of fight that could be used as recruit training, as was done with boxing in England and Judo in Japan. In the following decade, Mestre Sinhozinho (1891–1960) started teaching capoeira as a combat technique to middle-class youngsters in the upscale Ipanema neighborhood. Contrary to the approach taken by the military enthusiasts, Sinhozinho adopted capoeira as a pugilist art form and challenged other emerging martial artists, such as the Jujitsu Gracie brothers, to fight in a ring.

But how does one transform a practice that had been historically recognized as an infectious and morally denigrating social disease into a "safe," "honest," and "healthy" regimen of training that could lead modern Brazilian citizens to feel proud of their miscegenated heritage?

Generally speaking, despite their genuine enthusiasm towards capoeiragem, these elitist forerunners resorted to translating and systematizing this Afro-Brazilian modality of fight according to Eurocentric models of physical culture, sports, and military training. Implicitly connected to eugenics and the process of whitening, their efforts intentionally erased, or at least downplayed, capoeira's "undesired" association with blackness from their archives and repertoires. Meanwhile, the capoeiragem the populace played on the streets continued to be viewed as an "outlaw" activity that defied, at the same time, the capitalist work ethics, the positivist notion of order and progress, and scientific racism's understanding of "health" and "hygiene."

Later on, in 1928, the Italian Annibal Burlamaqui (a.k.a. Zuma) published a groundbreaking pamphlet entitled *Gymnástica Nacional (Capoeiragem) Methodizada e Regrada,* in which street capoeiragem is "distilled" into a modern combat sport with a training method and competition rules. Within Zuma's systematically described and illustrated list of kicks, one finds two particularly insightful maneuvers (*golpes*). The maneuver *pentear ou peneirar,* literally "to style (the hair) or to sift," makes a clear reference to the oscillating and offbeat *way* in which street capoeiras carried themselves. In his own words: "Throw arms and body in all directions with a swayed-action (ginga), so as to disturb the adversary and better prepare for the decisive strike."[31] Yet, in Burlamaqui's manual, the "sifting" maneuver, or gingado-walk, is set aside as a kind of "distracting transition" that one articulates ad hoc (without a clearly defined technique) between attacks. Conversely, Burlamaqui identifies "the guard" (*a guarda*) as the foundational position of capoeira. According to his detailed description, the proper execution of the guard requires a "noble and erect attitude." The capoeira pugilist first steps (the right or left foot) back with bent knees, transferring his bodily weight in that direction, while leaning the chest forward with a linear alignment of head and shoulders. His hands should remain open in a claw-like position. He then lifts his back heel, in order to launch an attack. Thus, by a stroke of genius, Burlamaqui's method incorporates the Afro-Brazilian swayed-walk or "sift" as a flourishing distraction, thereby maintaining a direct connection with the local capoeiragem practiced on the streets, while displacing its importance and replacing it with the "noble and erect attitude" of the guard.

Getulio Vargas

Getúlio Vargas' era (1930–45) set in motion a new paradigm that would radically change Brazilian society's understanding of its mixed-race heritage in general and of the ginga aesthetic in particular. As was outlined in the Introduction, the positive understanding of cultural miscegenation, first proposed by the sociologist Gilberto Freyre, is a crucial framework to this process. In sum, Freyre's theoretical conceptualization of cultural miscegenation or *mulatism* is versatile enough to untie old knots and tie new ones. It unties, for example, miscegenation from coloniality and its racial division of labor and ties it to culture. Then it unties ginga from Africanicity and primitivism, ties it to "Brazilianness," and calls it a "good thing" (as Martha Stewart would put it). In doing so, it offers a slippery yet insightful solution to the conundrum of how to construct a unifying image of Brazilianess without questioning the colonial matrix of power and the (unsolved) legacy of slavery/miscegenation.

After the coup d'état of 1937, under which President Getulio Vargas installed a populist dictatorship known as the "New State" (*Estado Novo,* 1937–45), his administration adopted a paternalist, and highly controlled, approach to the construction of an ethno-culturally inclusive nation-state. At one level, Vargas' populist dictatorship instituted blue-collar workers' employment rights such as paid vacation and sick leave, and created national institutes and programs to develop particular areas such as history, geography, performing arts, books, and folklore. At another level, Vargas' newly created Department of News and Propaganda sponsored (under strict regulation) nationalistic projects in the areas of radio-broadcasting, film and theatre, the printed press, tourism, and ancillary services, implemented alongside Rio de Janeiro's local network of industries of entertainment, all of which contributed strongly to the crystallization of a modern imagination of Brazil as an exotic (and docile) paradise of mixed peoples and cultures.

Departing from Freyre's understanding of cultural relativism, the Estado Novo's paternalistic measures fostered the emergence of a renewed *regime of intelligibility* (Rose, 1999), commonly known today as the "democracy of races" ideology. At its heart, this ideology incorporates non-hegemonic *cultural goods* (especially Afro-Brazilian products, processes, and customs) into a mixed yet unifying discourse of national identification that could be "intelligible" both locally and globally, while excluding the primary agents and producers of these ethno-cultural assets (i.e. Afro-Brazilians and their descendants) from these official narratives. Nevertheless, in practical terms, Brazil's African heritage continued to be regarded as a raw and/or primitive material reminiscent of the nation's colonial past, which ought to be tamed, distilled, or modernized so that it could be consumed by society at large and/or exported as a national good.

Little by little, the carefully constructed imagination of Brazil as a racially harmonious community *fixed* this once marginalized bodily technology connected to African heritage as an expression of Brazilianness. In the realm of performing arts, for instance, both the state and the local industry of entertainment tended to privilege singers, dancers, and performers whose "modern" performances incorporated non-European movement qualities particular to Brazil (e.g. samba's syncopated music or dance), but whose Caucasian phenotypes invariably catalyzed that society's acceptance and consumption of otherness. Some of the most emblematic entertainers, whose artistic career fitted this system of intelligibility like a velvet glove, include the radio singer and actress Carmen Miranda (see Chapter 5) and the ballerina Eros Volúsia (1914–2004; see Chapter 6).

Along with the celebration of samba as the national rhythm of Brazil, the wave of nationalistic euphoria framed by a paternalistic control of the Estado Novo played a significant role in the legalization of capoeira in 1937. First, the state used its "iron fist" to identify, regulate, and, in some cases, sponsor local "folk" traditions such as capoeira, which could safely engender an inclusive image of the nation-state. In addition, the local intelligentsia sought to rescue and celebrate this once "forbidden" practice through songs, poems, fictional novels, illustrations, photographs, and films. In this new scenario, capoeira's uncanny gingado-walk is reframed as an exotic trait unique of that miscegenated folklore. The ginga in capoeira is, subsequently, partially detached, or at least distanced, from the image of blackness and the shame of slavery. Meanwhile, the process of "folklorization" of capoeira rather reinforced its "difference" from (European) "High Culture." Over time, this "auto-exotic" way of articulating ideas corporeally, which was represented in "folkloric" form of expression (i.e. samba and capoeira) and was incorporated into European practices such as soccer, ballroom dance, and concert dance, became fixed as a sign of Brazil's uniqueness and therefore as a symbol of national identification.

Mestre Bimba

In addition to its recognition as folklore, the legalization of capoeira during Vargas' era led to the emergence of regulated training schools, where the practice gained wider legibility as a local form of gymnastic or fight modality, comparable to international methods of physical culture and martial arts such as boxing and Judo. A good example of this process is the "Center of Regional Physical Culture" (*Centro de Cultura Física Regional*), a capoeira school that Manuel dos Reis Machado (a.k.a. Mestre

Bimba) founded in 1937.[32] Mestre Bimba's teaching method played a crucial role, most importantly, in the transformative understanding of ginga in capoeira. Below, I offer a brief review of its development.

Born in Salvador in 1899, 11 years after the abolition of slavery, the Afro-Brazilian Mestre Bimba was the son of Luis Candido Machado, a locally famous fighter. In 1932, Bimba opened his first school of capoeira at Engenho Velho de Brotas, a low-income and predominantly black neighborhood in Salvador. Between 1932 and 1937, the school existed as a result of temporary permits that would expire and require renewal. He also taught capoeira to an army reserve center in Salvador (*Centro de Preparação de Oficiais de Reserva* (CPOR). According to scholars such as Rego (1968) and Assunção (2005), at the time Bimba was unsatisfied with the folkloric exhibitions of capoeira, especially because they "put too much emphasis on pantomime, and kicks were not efficient enough to face more serious challengers, especially those trained in the new martial arts coming from abroad" (Assunção, 2005, p. 132). Following the example of capoeiras of Rio de Janeiro, in 1936, Bimba challenged fighters from any martial arts modality to fight him in a pugilistic ring. After winning the match by a knockout, he sought to further "modernize" the practice by incorporating kicks and techniques from other (Asian and European) martial arts into his "regional" style of playing capoeira in a traditional circle or *roda de capoeira*.

In 1941, Bimba moved his capoeira school to the Pelourinho neighborhood at the heart of Salvador's historic downtown, near the prestigious Medical School of Bahia (*Faculdade de Medicina da Bahia*). The change of address had a dramatic effect inasmuch as it further facilitated the exposure of capoeira to members of the educated elites. The introduction of Bimba's *regional* method of physical culture to a specific group of white, upper-class college students was orchestrated by José Cismano Lima, then a medical student and eventually Bimba's right-hand man. In addition to learning capoeira, Lima and other medical students assisted Mestre Bimba in the legitimation of his school and the documentation of this training method according to the latest notions of "health" and "hygiene." Furthermore, with the enthusiastic support of his university students and their scientific vision, furthermore, Mestre Bimba's regional style was celebrated as a physical practice that could produce healthy bodies and disciplined minds, while connecting the (intellectual/white) youth to the folk/popular traditions of Brazil.[33]

Bimba's unprecedented teaching method, also known as the "ABC of Capoeira,"[34] established a basic routine of training to be repeated in order to achieve mastery, a hierarchical system of graduation with a formal ceremony, and a code of behavior inside the training academy, which instituted the obligatory use of uniforms, identification cards, attendance register, class timetable, didactic brochures, etc. In addition, incoming students had to perform a basic health test and show proof of employment in order to join the practice. Following in the footsteps of his predecessors in Rio de Janeiro (Burlamarqui, Sinhozinho, etc.), Bimba's training method embraced qualities such as efficiency, uniformity, accuracy, and productivity over apartness, playfulness, mockery, and leisure. Ultimately, under the ethos of "efficiency" of his method, capoeira became a bourgeoisie (paid) activity that took place inside a private establishment, with rules and a hierarchal order.

Bimba's transformative understanding of ginga can be found in the leaflet that accompanies his (now historical) LP *Curso de Capoeira Regional Mestre Bimba* (*Mestre Bimba's Course of Capoeira Regional*, J.S. Discos, 1960s). In this leaflet, ginga is isolated as a foundational "position," or unit of movement and is codified with the support of an analytical description, a geometric diagram, and an academic illustration. In the first lesson, one reads that:

> The gingado is the most important part of capoeira, the starting point of all other future acquisitions. It is the "fundamental" position of the capoeira player (capoerista), taken in a figurative sense, the key to agility and dislocation.
>
> The gingado is a back-and-forth movement, supported by a foot that stays behind, while slightly tilting the torso forward. The arms are raised forward at eye level, constantly protecting the face and torso. To exercise this movement, draw a square on the ground with chalk (fig. 1-D) and begin with the two feet on AB. Take the right foot (A) behind the left foot (D), and then return to the initial position. Next, take the left foot from B to C, and then return to B. This is the cycle that will be repeated indefinitely, maintaining the weight towards the tip of the toes and resting the body on the back foot, trying to avoid a very aggressive facial expression. Ask a friend to take you by the hand and practice the gingado as much as you can. Repeat, alone, the movements. (Bimba, n.d., 3, author's translation)

First and foremost, it is important to note that Bimba's method was geared towards teaching students to "get better in the shortest amount of time." Improvisation and apartness, as well as the undulation of the torso, are completely dismissed in favor of accuracy. Following this line of thought, the description above pays close attention to the movement of feet and arms, defining them according to geometric coordinates. Secondly, the foot pattern described above undeniably references, or rather departs from, Burlamarqui's description of "the guard," which had been published in the previous decade. In addition, similar to Burlamarqui's method, the efficiency of Bimba's methodology indirectly downplays the role of rhetorical strategies common to street capoeiragem, such as fakery and indefiniteness (*negaça*; see Chapter 4). In fact, as many scholars have pointed out, efficiency is the hallmark of Mestre Bimba's method. In addition, the illustration of "the gingado" in Bimba's leaflet features (as did the manual of his predecessor) erect and elongated men with Caucasian features.

Nevertheless, contrary to Burlamarqui's manual, Mestre Bimba's pioneering affirmation that the gingado is both "the starting point of all other future acquisitions" and "the key to agility and dislocation" strategically situates capoeira's sifting or swayed-walk as "the fundamental position of the capoeira player." Or, rather, it artfully combines the "erect alignment" of one's body, as suggested in Burlamarqui's codified "the guard" maneuver, with the swayed-walk of street capoeira. What interests us most is Bimba's description of "the gingado" as a codified unit of movement that should be learned before one can learn other attack and defense maneuvers. As listed on Wikipedia (see the Introduction), this description has since been adopted as the "standard way" of teaching capoeira in most styles.

While Mestre Bimba was invested in modernizing capoeira and producing efficient fighters, other senior players such as the members of the "Sport Center of Capoeira Angola" in Salvador were strongly concerned with preserving, and perhaps recuperating, the Africanicity of capoeira that, for many, was lost in Bimba's "ABC of capoeira." In Chapter 4, I will address the lineage of Mestre Pastinha, who is today recognized as the "father" of capoeira angola. For now, suffice it to say that capoeira angola constitutes one of the primary (although not the only) efforts to formally reinstate capoeira as an Afro-Brazilian self-defense practice. In this scenario, the ginga or gingado is understood as a shared, yet individualized, way of moving connected to African heritage. Ginga underlies all call-and-response maneuvers, while it projects black pride.

Conclusion: the ginga of Brazilian boys and girls

By the time the idealized "The Girl from Ipanema" took the world by storm in the 1960s, movements of bodily syncopation and/or undulation had already lost some of their stigma as a shameful "black thing" (*coisa de preto*). After Carmen Miranda's explosion in the USA on both Broadway and in Hollywood in the 1940s, for instance, ginga came to internationally epitomize a "typical" element of modern Brazil's celebrated cultural miscegenation rather than a fantasy of "black primitivism." In fact, going back to that bossa-nova song, one begins to realize that, following the footsteps of Miranda, the beauty and grace that Moraes idealizes bears little reference to local non-white bodies or the ideas commonly associated with the Afro-Brazilian socio-cultural context (e.g. slums, poverty, and lack of a "civilized" judgment of taste). As the lyrics indicate, "The Girl from Ipanema" describes a light-skined young woman walking by an upscale neighborhood on her way to the beach, where she most likely got her golden suntan. Detached from colonial/negative significations often attributed to non-whites (irrationality, hyper-sexuality, aggressively, etc.), the soft-spoken swaying action improvised by the idealized "tall and tan and young and lovely" girl from Ipanema exemplifies how, by that time, ginga had become a floating signifier for "Brazilianness." And, although that white-looking "girl" was imagined as an auto-exoticized object of desire, her behavior was considered a sexy "good thing."

Since the rise of the "Brazilian" brand of (exoticized) mulata beauty, bodily syncopation – now more closely attached to positive miscegenation than to negative blackness – has trickled down, bleeding across socio-economic demarcations. Though far from seamless, this positive reorientation gradually infiltrates both the international imagination of and about Brazil and the everyday life performances and speech acts (Austin, 1962) of a wide range of people born or residing in that geopolitical location.

After the Second World War, a progressive lift of moral censorship regulations led Rio's samba school parades to gradually incorporate partial nudity (e.g. breasts) and eventually feature young and good-looking women dancing samba in high heels and G-strings. Likewise, Rio de Janeiro's stage productions such as casino shows, burlesque musicals, and adult films adopted a similar pattern of exotic/erotic objectification in their scenes featuring sexy dancing mulatas, later known as "professional mulatas." In both these scenarios, which are further addressed in the next chapter, sexy dancing mulatas (of different shades) are progressively undressed, adorned with glitter, sequins, or feathers, and sold for mass consumption.

Part II
Analyzing Ginga

3
Understanding the Presence of Ginga in Samba Circles

The circle of the dance is a permissive circle: it protects and permits. At certain times on certain days, men and women come together at a given place and then, under the solemn eyes of the tribe, fling themselves into a seemingly unorganized pantomime which is in reality extremely systematic in which by various means – shakes of the head, bending of the spinal column, throwing of the whole body backwards – may be deciphered as in an open book the huge effort of a community to exorcise itself, to liberate itself, to explain itself. There are no limits – inside the circle.

Fantz Fanon, *The Wretched of the Earth*[1]

Introduction

In 1991, Rede Globo, Brazil's leading TV network,[2] launched *Globeleza*, a new division responsible for the coverage of carnival-related events. In the following year, the network released the first of a series of 30-second vignettes connecting the brand Globeleza, a synthetic play of words between *globo* (Earth's globe/Globo station) and *beleza* (beauty), to an exuberant dark-skinned samba dancer or *mulata*. Valéria Valenssa filled these shoes from 1992 to 2005, followed by Aline Aniceto (2006–13), Nayara Justino (2014), and Érika Moura (2015). Each summer, wearing nothing more than high heels and colored designs painted on her body, the mulata Globeleza re-emerges to usher Brazilian telespectators into the carnival season. She entices them to watch Globo's exclusive transmission of Rio's world-famous samba school parades.

In these eye-catching vignettes, created by the Austrian-born and naturalized-Brazilian designer Hans Donner, the female character is first shot dancing for the camera against a green screen background, which is later filled with computer-generated splashes of color in rainbow-like patterns. Donner's CGI innovations envelop Globeleza's solo performance with a futuristic stardust that dazzles viewers at home. Yet, there is nothing new about her dance. Her performance belongs to a long list of dances where women shake their hips to syncopated (i.e. black) rhythms. These exoticized objects of sexual desire had long been a "typical" image of or about Brazil. As further discussed in Chapter 5, Globo's characterization of the muse of the televised carnival follows in the footsteps of countless sexy dancing mulatas (*mulatas sestrosas*) featured in various staged performances. At the same time, it also references G-string feathered samba dancers "typical" of Brazil's infamous carnival parades.

Since 1992, the Globo network has aired a slightly new version of this scripted performance, from January until Fat Tuesday, all day long, between news reports, soap operas, and children's programs. And since Globo is by far the most watched TV channel in Brazil, Globeleza's upbeat and shameless dance for the camera has percolated into nearly every household in the country. It is one of the most widely known icons in the Brazilian media around carnival. Like Josephine Baker, Globeleza's athletic slender body, without large breasts or buttocks, contradicts the hyper-sexualized stereotype images of feminine blackness. Still, her vignettes reproduce a scenario under which (young) non-white female bodies are imagined as lascivious sources of endless pleasure, but are seldom treated with dignity or respect. The explicit eroticism of Globeleza's mass-mediated dance also places her bare flesh, and those she represents, at the inevitable position of submission or rape.[3] Without diminishing the physical beauty, technical skills, and charisma of the dancers who have portrayed the mulata Globeleza on TV, these exoticized/eroticized hip-shaking dancers continue to reiterate, through a violent process of control/domination/exploitation, the same set of complex and unresolved ideas established in the country since colonial times. The popularity of these vignettes in 2015 offers a clear example of how Brazil's pride-and-shame conundrum is yet to be solved. At the dawn of the twenty-first century, her dance reveals a particular kind of black femininity: one that is exotically appealing, sexually shameless, and widely available, yet not honorable.

In nearly all the videos, Globeleza stares at the camera, tempting the audience to follow its maneuvers across her lean, brown, vibrant, dancing body. Always in motion, she shuffles her high-heeled feet as she spins around her axis or waves her arms around her torso. With a hypnotizing smile, this sexy dancing mulata expresses nothing but joy, as Jorge Aragão's theme song plays in the background:

> Come / to be happy / I'm on air with the Globeleza / I'm having a real good time.
> [*Vem / pra ser feliz / Eu tou no ar com a Globeleza / Eu tô que tô legal.*] (Author's translation)

As the song progresses, the camera either zooms in and out or revolves around her sensual body, filling the screen with her toned and flawless legs, breasts, belly, crotch, and buttocks – highly desirable pieces of flesh coated in glitter. In sync with the visual interplay across (masculine) gaze, (feminine) seduction, and (objectifying) desire, the syncopated song clinches an enduring double entendre:

> On the TV screen / in the middle of this crowd / we will see each other at Globo.
> [*Na tela da TV / no meio desse povo / a gente vai se ver na Globo.*] (Author's translation)

In this chapter I move away from this fixed idea of samba as an exotic/erotic dance performed by dark and half-naked women framed into virtual "scenes of subjection" (Hartman, 1997). Instead, I investigate samba dancing as a movement practice connected to the African heritage in Brazil. My exploration below is not aimed towards translating samba as a set dance form, with a codified set of rules and a fixed point of origin. The movement practice I discuss in this chapter should be understood instead as a flexible, improvisational, and highly personalized way of moving, historically cultivated in Brazil. Rather than a product, samba dancing is a process. In this regard, I consider it important to distinguish samba as a participatory social dance, examined here, from its appropriation and/or adaptation in folkloric demonstrations, concert dances, dance videos, or films. While in all these scenarios the choreography may be improvised, in social dance settings there is no clear distinction between performers and audience.

In this chapter I focus on samba practiced in a circle, or *samba de roda*, a sub-genre that precedes other forms such as samba parades and ballroom samba. In particular, my movement analysis offers an in-depth examination of one of the foundational ways in which samba dancers: (a) discipline their body to articulate syncopated dialogues across bodily parts, especially between the hips and feet; (b) structure their choreographic call-and-response improvisations inside a human circle; and (c) dialogue with other dancers and musicians at the rim.

In order to accomplish that, a few considerations must be established. First and foremost, samba should be understood as an umbrella term that encompasses a wide range of inter-related dancing events, musical rhythms, choreographic patterns, lexicons of movements, costumes, and so forth. Secondly, though particular kinds of samba dancing adhere to specific musical genres, the genealogy of these rhythmic mixtures often differs from the genealogy of the movement vocabulary they accompany.[4] Thirdly, while samba should be regarded as a transatlantic and transcultural form, it retains strong connections to African – especially Bantu – cultural matrixes.

At the macro level, samba dances have incorporated, a range of external influences, starting with the use of the Portuguese language, the Moorish *pandeiro* (a type of tambourine), and, later on, the Amerindian traditions of feathered costumes (in samba parades). In addition, as people migrated from one region to another, from rural plantations to port cities and urban centers, these local black dances continued to absorb new elements (e.g. movement vocabularies), assuming different names and functions. Yet, as this chapter seeks to clarify, both samba dancers and musicians employ the ginga aesthetic as an underlying logic with which to construct their transcultural retentions and ruptures, or swing between meters. In samba, I argue, both dancers and musicians must have a kinesthetic understanding of the ginga aesthetic in order to (re)produce or dialogue with such rhythmic patterns. The ginga aesthetic functions as an organizational logic through which people think and act in samba.[5] From this system of organization, people may choose to articulate the concept of tonal syncopation through movements and/or sounds.

As Sodré (1998) synthetically summarizes, percussive samba music is the result of an interaction between bodily movements and instruments. Similar to the work of samba musicians, samba dancers deploy a variety of physical techniques to instantiate the frictional principle of (tonal) syncopation corporeally. In 1878, the novelist Machado de Assis provided us with a vivid picture of this interaction. In the short story entitled "O Machete,"[6] the narrator describes a musician playing an acoustic guitar as follows:

> His entirety accompanied the gradation and variation of the notes; he would incline over the instrument, stiffen the body, he would tilt his head, at times to one side, other times to the other, he would

> raise one leg, smile, melt his eyes or close them at places that it seemed emphatic. Listening to him play was the least, watching him was the most. Those who only heard him, could not understand him. (Quoted in Wisnik, 2004, p. 22, author's translation)[7]

The excerpt above illustrates one of the ways in which musicians have used their embodied awareness of polycentrism/polyrhythms in order to reproduce syncopated soundbites. Upon closer examination of Machado's description, one comes to understand that the different parts of the musician's body *move with* or *against* his instrument, as he fingers syncopated rhythms with his guitar. In other words, while the entanglement between the hands and the guitar's strings are engaged in making them audible to others, the entire body vivifies the kinesthetic understanding of tonal syncopation.

At the same time, samba dancing is not just a *reaction* or *response* to samba music, although dancers vivify syncopated rhythms, making them visible. Regarding the relationship between the polyrhythms that both dancers and musicians produce in samba, dance scholar Barbara Browning confirms that "the dancer may make reference to all of them with different parts of her body" (1995, p. 14). In samba dancing, different parts of the body participate in the construction of multi-meter arrangements by either following the audible score or by layering new rhythmic patterns above the main riff. While musicians play with musical instruments, dancers produce rhythmic patterns by "playing" with their own isolated body parts in dialogue with the ground/gravity. As a result, both musicians and dancers produce syncopated rhythms, which bounce back and forth, through call-and-response interactions between seeing and hearing.

In Brazil, this way of disciplining bodies to move has contributed to recuperate-cum-invent Africanist knowledges that clearly diverged from Europeanist expectations. Blacks and mulattoes have often used it as an embodied strategy of reaffirmation of their singularity, against the epistemic violence that resulted from the colonial encounter. Paraphrasing Fanon's quote above, inside protective and permissive samba circles across Brazil, the ginga aesthetic has systemically structured the efforts of marginalized communities *to exorcise themselves, to liberate themselves, and to explain themselves.* Besides, one should bear in mind that the aesthetic values and procedures commonly associated with ginga have been cultivated within territorialized black spaces since the institution of the African slave trade, often within the context of religious ceremonies (see Chapter 2). This is not to say, however, that modern samba is a profaned version of a previously existing sacred form. Rather, it testifies to the significant role that these black religious events had in the: (a) recuperation-cum-invention of Africanist aesthetic knowledges and cultural memories in the diaspora; (b) the re-creation of a pliable place (or cypher) where participants could re-link their lived cultural heritage to affects such as pleasure, happiness, honor, and pride; and (c) the delineation of performative territories where they could *speak for* themselves. As I see it, the social events organized by people initiated in these Afro-Brazilian religions reiterate, or *evoke*, aesthetic aspects of the embodied system of beliefs/cosmology that are called on, or *invoked*, during their religious, often secluded, ceremonies of incorporation and devotional dances.

The *Irmandade da Nossa Senhora da Boa Morte* (literally the Afro-Catholic Sisterhood of Our Lady of the Good Death) located in the once prosperous port city of Cachoeira, in the countryside of Bahia, offers a good example of this argument. At one level, much of the Sisterhood of Boa Morte's fame derives from its rare position inside Brazil's patriarchal and racist society as a long-lived religious and mutual aid organization composed exclusively of elderly female descendants of enslaved Africans and connected to the Candomblé religion. On another level, the Sisterhood's history is deeply intertwined with the cultivation of samba music and dancing in that region as well as its dissemination to the rest of the country. Each August, for example, the Sisterhood of Boa Morte sponsors its Afro-Catholic celebration, punctuated by traditional samba circles and an array of African-inspired dishes (votive food).

The cultural activities of two distinguished members of Boa Morte, in particular, illustrates the extent to which this Afro-Brazilian religious organization has informed the trajectories of samba in Brazil. Tia Ciata (Hilária Batista de Almeida (1854–1924), who is further discussed in Chapter 5) pioneered the organization of modern black street parades during Rio de Janeiro's carnival. Her now-famous house dancing parties are also considered to be an important breeding ground of samba as a musical genre in that city. Born a few years after Ciata's death, Dona Dalva (Dalva Damiana de Freitas, b. 1927) is a composer, singer, and dancer whose house in Cachoeira, the *Casa do Samba de Dona Dalva*, operates today as a cultural center with weekly samba events. In parallel to her religious activities at the Sisterhood of Boa Morte, since 1958 D. Dalva has spearheaded the acclaimed *Grupo de Samba de Roda Suerdieck*.[8] Her wisdom and experience with samba circles practiced in the countryside of Bahia and her historical link to Boa Morte also played a fundamental role in the laborious process that resulted in UNESCO's recognition of this regional practice as an intangible cultural heritage of humanity in 2005 (see Conclusion).

Departing from these premises, in the next few pages I will examine how samba dancers' cultivation of the ginga aesthetic enables them to improvise non-hegemonic discourses of identification from a place of otherness. Above all, this chapter seeks to demonstrate how samba dancing, as a lived experience, defies the traumas of colonization. Dancers create improvised choreographies inside the circle, for instance, mixing physical pleasure with cultural agency. When they dance, they choose from a pull of steps, efforts, pathways, and rhetorical strategies they have learned over the years in order to stand apart and have fun. When done well, dancers transform their transient interactions into memorable discourses. The continual reiteration of this pleasurable way of dancing during social gatherings has left undeletable traces within the memory of dancers and viewers alike, despite its ephemeral duration. As the Bahian counter-culture artists Caetano Veloso and Gilberto Gil sing on their joint album *Tropicália 2* (1994), samba is both the "father of pleasure" and the "son of pain."[9] Drawing a circular line around samba and the Afro-Brazilian concept *axé*, or the "power-to-make-things-happen" (Thompson, 1983, p. 5), Veloso and Gil proclaim this way of moving as "the great transformative force." In a country built upon the widespread trafficking of human flesh, torture, and abandonment, I propose that the power of embodied knowledges grounded in African heritage that are articulated through festive and playful acts, especially through physical pleasure, should not be underestimated.

A final disclaimer should be given at this point. As stated above, the primary goal of this chapter is to investigate the presence of the ginga aesthetic in samba de roda, paying close attention to the role and function of this polycentric and polyrhythmic way of moving within the context of improvisational dancing circles and the effects it has produced in that particular setting. There are a number of issues that, however important, are beyond the scope of this chapter and therefore will not be fully explored below. They include: (a) the various kinds of samba dances and their related musical rhythms; (b) the lyrics and rhythms of samba music; (c) the rivalry between Rio's and Bahia's samba styles or the primacy of one over the other; (d) the many stereotypes associated with samba-like music, dance, and events since colonial times, including the *malandro* (a type of street-smart dandy); (e) other syncopated black dances cultivated across Brazil, which may arguably have influenced modern samba dancing, such as *jongo, afoxé,* or *frevo*; (f) the emergence of new rhythms related to samba after the

Second World War, including *bossa nova, samba-reggae, samba-rock,* and *pagode.* I will make reference to some of these elements and issues inasmuch as they support my main argument.

The "pre-history" of samba

In *The Mystery of Samba* (1999), the ethnomusicologist Hermano Vianna argues that samba as we know it today should be understood as an "invented tradition" orchestrated during the first decades of the twentieth century. Vianna proposes that the crystallization of samba as a distinct musical genre within Rio de Janeiro and its subsequent branding as Brazil's national dancing rhythm during Vargas' era (1930–45) was a carefully constructed project. This is not to say that samba was an unfamiliar term on the street of Brazil. Nevertheless, as Vianna puts it, "there never existed a well-defined, 'authentic' samba genre prior to its elaboration as a national music" (1999, p. 112). The "invention" of samba and its elevation as a national symbol, he explains, involved a series of transversal connections across working-class artists (e.g. musicians, composers, interpreters, and dancers) of different ethnicities; elitist intellectuals (e.g. social scientists, journalists, directors, and playwrights); owners and managers of mass-media industries (e.g. newspapers, recording studios, and radio stations), local and national politicians, and a few international figures (Vianna, 1999, p. 112). These different focus groups collaborated, in particular, to incorporate Rio de Janeiro's local black culture into a unified discourse of national identification. Or, as Antonio Herculano Lopes puts it, the construction and mass-dissemination of samba emerged from dialogues and confrontations between Rio's European-aspiring elites, who controlled the means of communication almost exclusively, and the populace, whose cultural traditions was sharply influenced by Afro-Brazilian heritage (2000, p. 29). As a result, Vianna ponders, Rio's local black community has played a marginalized, although significant, role in the construction of the image of Brazil as the Kingdom of Samba (1999, p. 92).

In order to understand how samba, of all the rhythms and dances available in Brazil, is chosen to symbolize the nation, below I provide a brief retrospective of the "pre-history" of samba, starting from the "birth" of this genre in Rio de Janeiro and moving backward in time to retrace what happened before its inception. My goal here is to bridge the gap between a range of colonial black dancing styles and rhythms associated with the umbrella term "samba" since colonial times and the constructed imagination of Brazil as a samba nation in the age of mass-reproduction. The city of Rio de Janeiro, one should know, had functioned as a kind of "cosmopolitan postcard of Brazil" – or synthesis of nationality – amplified via mass-culture since the nineteenth century.[10] Also, my backwards trajectory in that city takes into consideration the debates currently available within the scholarship on samba, including the already mentioned "originality" of samba as a modern tradition and the "authenticity" of samba as a Kongo-Angola tradition transplanted to and re-created in colonial Brazil, which is supported by scholars such as Nei Lopes (1992). In doing so, I will offer an insight on the genealogy of ginga in samba dancing. In Chapter 5, I will return to the topic of samba via Carmen Miranda in order to further deconstruct the notion of samba as a fixed idea connected to carnival, Brazil's "national" holiday, and local stereotypes of non-white females.

Our journey begins in 1917, when Odeon Records released the historic "Pelo Telefone" ("By Telephone"), the first recorded song to mention the word "samba." Though technically considered a mixture of maxixe and samba or "maxixied samba" (*samba amaxixado*), this carnival hit-song symbolically marked the "official birth" of samba music as a cultural commodity. Connected to Rio's working-class pageant-parades during Mardi Gras, the twentieth-century maxixied samba combined the marked pace of a march with other hybrid black rhythms such as the maxixe and the *lundu* (discussed below). Meanwhile, the choreographic structure of these early modern black carnival parades was molded after the proto-Catholic colonial pageant-parades sponsored by black-only Irmandades such as *Congadas*. These performances acted as ambulatory dances similar to New Orleans' second line dancing, where revelers move in linear and unidirectional trajectories across streets and open spaces, dancing either guided or followed by a percussive marching band. Nevertheless, prior to the formal institutionalization of "samba schools" in the 1930s, linear and circular styles of samba dancing would often overlap on the streets of Rio's carnival. Across peripheral neighborhoods such as Saúde and Cidade Nova , for instance, it was not uncommon to see a pack of (largely non-white) dancers parading to syncopated rhythms, wiggling their hips or sending their flexible spines into playful undulations as they paraded. At strategic points, such as in front of the house of distinguished samba personalities (e.g. Tia Ciata), these krewes (*ranchos*) would break their linear structure to form a dancing circle. After a while, the pack would reconvene into their ambulatory formation, giving continuation to their linear evolution.

As these flexible interactions are taking place, the industry of entertainment disseminate fixed narratives connecting Rio's local black dances and rhythms to stereotypical characters such as sexy mulatas and malandros, mass-disseminated via records, radio programs, casino shows, staged musicals, motion pictures, and specialized magazines. For A.H. Lopes, these fixed narratives disseminated by the local entertainment industries were influenced by the dialogue between the dominant culture, which followed the Northern hemisphere's standards, and the so-called "popular culture" centered on Afro-Brazilian heritage (2000, p. 29). Little by little, the image of culturally mixed *cariocas* (locals of Rio de Janeiro), with their (imagined) bodily swing, refined musicality, and aptitude for improvisation, is adopted as a synthesis of Brazil's national character or identity.

As will be further addressed in Chapters 5 and 7, the maxixe rhythm latent in "Pelo Telefone" dates further back to the second part of the nineteenth century. To be precise, initially maxixe was not considered a distinct genre of music or dance, but rather a lascivious (black) "way of dancing" to popular ballroom rhythms, such as polka, tango, and habanera. Named after *burr gherkins*, at the time the cheapest vegetable on the market,[11] maxixe dancing was practiced at establishments of "questionable reputation" located in working-class neighborhoods. It was initially despised by the European-aspiring elites of Brazil. Yet, in a turn of events similar to what happened to the Argentinean tango (Savigliano, 1995), soon after distinguished maxixe dancers made their debut in Paris balls around the 1910s, this now properly "tamed" and codified dance became an auto-exotic sensation amongst the local elites in Brazil. Subsequently, maxixe spilled into populist musical theatre productions, whereby composers such as Chiquinha Gonzaga and Ernesto Nazareth orchestrated hybrid compositions for piano, mixing the Eurocentric metric of polka with elements of previously existing black rhythms, such as lundu and *chula*, a type of music played in samba circles.[12] Within music sheets, these innovative hybrid "fusions" were archived as polka-lundu, Brazilian tango, and (later) maxixe music.

Choreographically speaking, maxixe dancing – and its twentieth-century successors *samba de gafieira*, *forró*, and *lambada* – may be understood as a transcultural style, which combines the Afro-centric ginga aesthetic with a Eurocentric organization of social bodies in space (i.e. partner dancing). In all these dance styles, the distance between the dancing pair is practically erased, allowing dancers not only to move together across space, but to move in and out of each other's private spheres. With their hips and thighs pressed against one another, for example, maxixe dancers tend to push and pull their weight horizontally, curving their upper bodies over and under each other. This Afro-centric way of dancing in pairs also pushes dancers to constantly swing, shuffle, or evade their upward posture in order to compensate for the constant feet shuffling and the dynamic and offbeat dislocation of their closely connected pelvis that moves independently of their upper body. The result is a "passionate" abandonment of Eurocentric uprightness and decorum. Invading each other's privacy with sexually charged frictions, maxixe bends all embodied codes of morality, civility, and social etiquette disseminated through the Western codes and regulations of ballroom dancing. As Jota Efegê notes, maxixe dancing was long condemned as an "excommunicated dance" or the "the dance that inebriates," until the late 1930s (1974, p. 46).[13]

As an example of how this hybridized ballroom form was closely related to the samba danced within a circle in the context of carnival, Chasteen comments that, at the turn of the twentieth century in Rio de Janeiro: "The main difference in the newspaper reporters' usage of the words maxixe and samba lay not in the movement but choreography. A man and a woman who faced each other inside a circle were dancing samba. But if the couple moved into close embrace, they were dancing maxixe" (2004, p. 45). In other words, these two forms may be understood as homologous ways of imagining and organizing the dancing body, with distinct ways of structuring the interactions between dancers in social space.[14]

Moving further back in time, one finds the lundu, one of the first Afro-Brazilian dancing rhythms to be included within Rio de Janeiro elitist's archives (e.g. musical sheets, novels, paintings, and prints). Lundu – and, more specifically, "danced lundu" (*lundu bailado* or *lundu-dança*) – emerges at the dawn of nineteenth-century Rio de Janeiro, after the Portuguese court moved to Brazil as a consequence of the Napoleonic Wars. Nei Lopes (1992) also traces a rural form of danced lundu strongly connected to Afro-Brazilian circular dances cultivated in plantations and maroon communities in the states of Bahia and São Paulo. Though lundu was a mixture, according to Tinhorão (1988a and 2000) and Nei Lopes (1988, 1992, and 2005), it is strongly connected to practices traditional of Central Africa (especially the Kongo-Angola region). From the available visual and textual archives, I envision the danced lundu at the crossroad of earlier mixed folk dances transplanted from Europe, such as the *fofa* (Portugal) and the fandango (Spain), African dances from the Kongo-Angolan region, and a few other colonial Afro-Brazilian dances, such as the *samba de umbigada* (addressed below).

There is little descriptive information on lundu's lexicon of movements, but it is clear that its insinuating use of the hips was considered indecent at that time and thus not proper to "honest" women. Perhaps the most scandalous step was the *umbigada*, literally "navel greeting." Briefly, dancers "greet" or "challenge" one another as they trade places inside the circle by pushing their hips forward and bumping their navel area in the air, hence the name. Across Imperial Brazil (1822–89), the popularity of lundu increased, although it is unlikely that the European-aspiring upper class danced lundu as the working-class blacks and mulattoes did. Through the commercialization of music sheets, lundu became further incorporated into mainstream social activities such as theatrical productions, court balls, tea parties, and piano lessons for "honest" women. In this process of transculturation and recontextualization, the "refinement" of lundu dancing in upper-class circles partially disassociated it from the irrationality or primitivism attributed to the so-called "black dances" in general and, in particular, the lasciviousness or shamefulness allocated to non-white dancers, especially women well versed in the ins-and-outs of the "navel greeting."

Prior to danced lundu, one finds a variety of syncopated circular dances linked to Kongo-Angolan batuques, many of which employed "lascivious" movements involving tapping of the feet, shaking of the hips, bending of the spine, and the above-mentioned umbigada. The ethnographer/folklorist Carneiro (1961) was the first to group these various dances under the name *sambas de umbigada*. Yet, today only a few regional manifestations such as Rio de Janeiro's *jongo* and Maranhão's *tambor de criola* continue to articulate the "navel greeting" inside the dancing circle. The individualized hip-and-feet "agitation" is currently regarded as the common denominator, shared across these circular dances.

Shall we samba dance?

Samba de roda literally means samba practiced in a circle. However, the use of this expression is historically attached to Bahia, one of the most prosperous states in colonial Brazil. Similar to street dance cyphers in the Americas, these performative spheres create a "domestic" or "familiar" environment that provides entertainment to the community, while cultivating aesthetic knowledges and transmitting cultural memory associated with their African heritage. Samba circles have functioned thus, as territorialized black spaces conducive to the exchange of experiences coming from different places, promoting both tradition and innovation. Furthermore, as productive places, these participative arenas foster playful exercises of creation, including the call-and-response improvisation of rhymes, rhythms, and movements, all of which preserve a systematic yet non-hegemonic way of organizing bodies to move and make sense of their reality. As previously indicated, samba circles act as protective and permissive spheres through which a community exorcises itself, liberates itself, and explains itself (Fanon, 1963, p. 57).

Regarding the cultural recuperation-cum-invention that takes place in these social events, Moura (2004) proposes that:

> Every samba circle is unique and irreproducible ... Like any ritual, the circle preserves and updates what was is in its origin. There, the tradition dialogues with the present in the course of history. Everything happens from the available material conditions, but it is essential to respect the fundaments. That is, the participants do not expect ideal conditions to act, but never act contrary to the canons established by the community – because the world of ritual and "totally relative to what happens in the quotidian life" (Roberto da Matta) ... As a dialectic between the everyday and the utopia, the samba circle establishes in the practitioner the illusion of eternity, as if time had stood still and the world stood apart, turning the circle into an enchanted "rest stop for the warrior." (Moura, 2004, p. 23, author's translation)

As Moura implies in the above quote, samba circles act as pliable frameworks where anyone (dancers, players, singers, etc.) may cite, appropriate, and recontextualize shared principles with personal opinions, constantly re-inventing tradition. Furthermore, Moura concludes that samba circles are creative spaces that generate songs, rhythms, movements, social interactions, and a sense of community (2004, p. 31). In other words, these cultural events produce and therefore precede samba music and dancing, and not the other way around. To paraphrase the historian J.J. Reis, the kinesthetic imagination that dancers (and musicians) articulate at samba events, "transform memory into lived culture" (2005, p. 210). Within these post-colonial circles, participants patch together ideas and motifs extracted from different places, (past) experiences, and (future) goals. In this protective and permissive context, they enact choreographed discourses that also recuperate and/or reconceptualize their sense of selfhood. And last, but not least, samba circles are festive sites where people from different ethnicities, classes, and creeds gather to sing, play, dance, laugh, have fun, and, above all, see and be seen.

Along with the wide variety of rhythms, instruments, songs, accents, and verbal expressions that participants employ in these jam sessions, samba dancing offers, in particular, an embodied way to articulate non-hegemonic discourses of identification that connect blackness to pleasure, grace, and dignity. Despite change and rupture, the efficacy of these improvised performances of creation, erasure, transference, and substitution relies on the participant's ability to affirm their alterity with positive values and symbols. Yet, while samba music has been the subject of a wide range of scholarly studies, thus far, all attempts to analyze samba de roda's dancing endeavors have failed to move beyond the cataloging of units of movements and footwork patterns. Perhaps due to its improvisational character, attempts to identify and critically analyze the underlying logic structuring their "seemingly unorganized pantomime," as Fanon puts it, is practically non-existent. In order to overcome such a gap, below I will consider how samba dancers employ the ginga aesthetic inside the circle, in order to construct syncopated conversations between bodily parts, dancing bodies, dancers and musicians, or performers and audience. Because samba dancing privileges apartness in movement (style) over uniformity and precision (form), my goal is not to codify all the possible movements, steps, and gestures associated with samba circles. Instead, in my analysis, I examine how polycentric and polyrhythmic dancing bodies adhere to a particular set of qualities of movement, tactics, strategies, and pathways in order to remain connected to the whole while standing apart.

Mapping the presence of ginga in samba de roda

In samba circles, the syncopated conversation between shuffling feet and wobbling hips, commonly known as *samba no pé* (literally "samba in the foot"), acts as departure point from which all other movements develop. This is how polycentric and polyrhythmic bodies generate the ginga in samba dancing. First of all, the feet delineate a particular shuffling pattern that may be described as follows: (1) foot A slides back; (2) foot B slides or skips forward; (3) foot A steps in place; and (4) foot B slides or steps back, switching places with the other foot. This back-and-forth shuffling of the feet marks the basic 2/4 beat, where the tonal steps (TA, ta, TA, ta) are accentuated by the back foot (A). Meanwhile, the atonal steps (TA, ta, TA, ta) are lightly marked by the other foot (B), which first slides forward then shuffles back, with a slight outward rotation of the leg. As foot B falls back to restart a new riff (TA), foot A slides or skips forward, hitting the next (atonal) beat (ta).

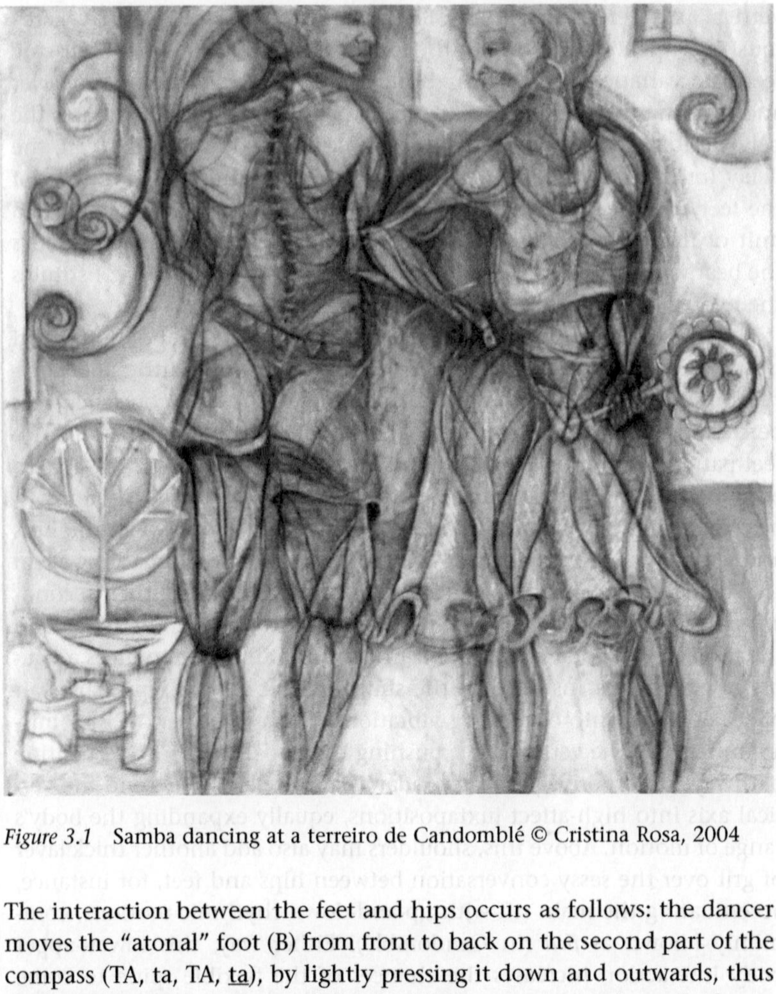

Figure 3.1 Samba dancing at a terreiro de Candomblé © Cristina Rosa, 2004

The interaction between the feet and hips occurs as follows: the dancer moves the "atonal" foot (B) from front to back on the second part of the compass (TA, ta, TA, ta), by lightly pressing it down and outwards, thus provoking a slight rotation of that leg all the way up to where the femoral head meets the hip socket (acetabulum). In this fast-paced contratempo, the outward rotation of the front foot triggers the hips to snap or slide sideways, then bounce back once that foot falls back to "catch" the next riff. More importantly, as the footwork (re)produces the 2/4 beat, the hips' jigsaw oscillation "lags" slightly behind that metric, thus performing momentary dislocations in time. As a result, the pelvis jiggles right after the rotation, somewhere between the two beats, leaving a gritty trace in the air. Like an "afterthought," the messy wobbling of the hips further provokes a gentle undulation or reverberation of the flexible vertical axis (vertebral column). By the same token, while the navel area must be "open" to respond to the footwork's call, it may also "call" the feet into variations on this rhythmic theme by contracting its muscles and gyrating the sacrum in a "figure of eight." In the end, whether the pelvis initiates the movement or responds to the lower limbs' call, the delay (or dislocation in time) between the back-and-forth shuffling of the feet and the lateral agitation of the hips (re)produces a syncopated unit of movement. This offbeat oscillation, where the footwork keeps the beat while the hips wobble in and out of place and time, constitutes the basic ginga of samba.

This hips-and-feet dialogue, *samba no pé,* acts as a foundational grid from which dancers engage other parts of their polycentric and polyrhythmic bodies to improvise variations and, eventually, construct their personalized way of dancing samba. In other words, once this hip-and-feet pattern of motion is understood kinesthetically, dancers are able to employ the ginga aesthetic to articulate other kinds of frictional and offbeat dialogues. For instance, the initial conversation between the hips and feet may reverberate across the dancers' upper body, enabling them to articulate their flexible vertical axis (vertebral column), their secondary pairs of scales (shoulders), their upper limbs (arms and hands), and their head. This total engagement from the ground up produces a couple of effects. For instance, as the shuffling feet send the primary pair of scales (hips) into oscillating vibrations, it may also set off the "misalignment" of the vertical axis, pushing the upper torso into serpentine pathways. Alternatively, the secondary pair of scales may send the vertical axis into high-affect juxtapositions, equally expanding the body's range of motion. Above this, shoulders may also add another thick layer of grit over the sassy conversation between hips and feet, for instance, accentuating the main riff with up-and-down rhythmic motions, commonly known as *jicá* (in Candomblé). Either way, this lower–upper body interaction increases the chances of the flexible spine to come in and out of balance and alignment. As the secondary pair of scales starts to dialogue with the lower body, they may also drag the upper limbs along into that auxiliary conversation. In turn, the arms and hands may articulate free-flowing accents, abstract flourishes, symbolic gestures, or pantomime, all of which may reverberate the hips-and-feet interaction or add new layers to that primary conversation. Combined, these decentralized and multi-meter interactions across different parts of the dancer's body produce an infinite number of playful combinations, often designed to further capture the imagination of those at the perimeter of the circle.

Dynamics

Following my commitment to explain how samba dancing may be regarded as a process grounded in improvisation rather than a codified dance practiced under a strict set of rules and protocols, below I will discuss the dynamics that the ginga aesthetic enables dancers to articulate or put into practice within a samba jam session. As a rule of thumb, dancers tend to structure their qualities of movement following a call-and-response arrangement or syntax. Therefore, I will refer to them in pairs. One of the most common interactions in samba de roda pertains to bodily *reverberation* and *isolation*. In order to increase the interaction with the spine, for instance, a dancer may contract and release the muscles within the navel area, while swinging the hips. This active push-and-pull of core muscles, as well as the rotation of the hip joint, calls (or push) the tailbone to wobble back and forth in space. This rippling effect enables a pliable articulation of the polycentric core in either seamless or abrupt compositions. Conversely, one may constrain the movement of the upper body, thus drawing attention to the isolated movements of hips-and-feet syncopation. Another common kind of variation pertains to *break* and *flow*. Dancers may either sustain a constant sense of agitation and fluidity or rush into sudden shifts, infusing their fast-paced oscillations with unexpected effects of contrariety, for instance, juxtaposing heated footwork and serene expression. Dancers may also break the flow of motion by momentarily "freezing" or "squatting" abruptly and getting up again. Spiral *folding* and *unfolding* is also a very popular pattern of movement. "Screw" (*parafuso*), for example, is a common unit of movement employed to include variations in pace and levels. Briefly, dancers cross their legs and spin while descending to or ascending from the ground. In addition, dancers often articulate *dragging* and *skipping* dynamic. The *miudinho*, a particular kind of short and fast-paced footstep with which dancers drag their feet on the ground, is a simple yet effective way of calling the hips to wobble in short-length frequency waves. Also, depending on the type of shoes a dancer may be wearing, such motion also adds new sounds to their rhythmic composition. Conversely, skipping footwork adds new lively textures to the already-syncopated samba beats. Once dancers learn how to employ ginga to articulate these and other kinds of movement qualities, they are able to step into the center of the circle to dialogue with external elements, such as the musical rhythms, the surrounding space, other dancers, and the participant audience circumscribing the dancing space. Despite the high level of permissiveness and improvisation, more often than not, participants adhere to a series of "unspoken" rules, learned through observation and reinforced by senior dancers. These guiding protocols orient, for instance, when to enter or exit the central arena and how to greet one another. In general, these samba jam sessions adhere to the following sequence of events. Once participants gather and form a human circle and the musicians establish the rhythm, the leading singer introduces a call-and-response song. Afterwards, the dancers enter the circle to improvise, either one by one or in pairs. After a while, either a new dancer steps into the circle to join or replace the previous one or the central dancer moves towards the circumference and "chooses" a partner or replacement. The time span between rotations may be determined by a series of subjective factors, such as the ability of a dancer to please the crowd or the timing with which a challenger foresees an opening to "break" the performance of the previous dancer. Generally speaking, each solo or duet combines rehearsed and improvised actions, arranged in a call-and-response fashion, which are executed in relation to the syncopated music. Dancers at the center may further dialogue or interact with the participants at the rim.

Call-and-response interactions in samba de roda

In order to understand how the concept of call-and-response works in samba circles, here I transcribe an initial pattern of interaction between two dancers based on my extensive experience participating in and observing such improvised enactments. Initially, the person dancing in the center approaches the edge of the samba circle with a "miudinho" step and makes eye contact with another dancer. Instead of the old-fashioned "navel greeting," she might open her arms and tease her with an inviting hip-swing, as if asking "Do want to join me?" or "Can you synchronize with my riff?" In response, the solicited dancer may step into the circle and begin to dance, as if responding "Sure, here I go..." Placing her hands on her waist (considered a sign of confrontation), the upcoming dancer may propose a variation of samba no pé tinted with an eloquent thrust and an arm variation. Staring at the inviting dancer, her movements may inquire "How would you answer this?" or "Can you sync with that?" In response, the other dancer may also place her hands on her waist, accepting the challenge. Once they synchronize their steps, she may propose a new break-and-flow pattern of hip-syncopation, this time spinning around her vertical axis with virtuoso sensuality, as if saying "How about that?" At this point, the challenged dancer may distance herself to delineate a wide and curvilinear pathway alongside the rim of the circle, using the same *miudinho* steps, in order to warm up the audience towards her wobbling hips. If the audience's response is positive, the first dancer may choose to step off the center and leave room for this new dancer to improvise alone. In this hypothetical sequence of actions, each call and each answer pushes the conversation one step further towards one (or more) direction(s). Nevertheless, in reality, samba dancers employ a much wider range of tactics and strategies in order to string together dramatic inquiries, propose metaphoric questions, play out improvised answers, or simply unravel riddles.

Tactics of persuasion: the role of ginga and the effects it produces in samba

As previously stated, departing from the ginga aesthetic, the frictional dialogues that samba dancers mobilize between the hips and feet function as the foundational mechanism from which they "juggle" their center of gravity/weight back and forth across different parts of their bodies. Similarly, inside the circle, samba dancers employ this polycentric and polyrhythmic logic of interaction, or "hip-play" (*jogo de cintura*), to oscillate between their commitment towards being in sync with the other(s) and standing apart. Senior dancers are capable of improvising complex duets, for instance, in which they engage with their partners, while showcasing their own eloquence.

Combining pleasure and derision, or malleability and fierceness, the dancers seek to articulate non-verbal dialogues filled with libidinal double entendres, tongue-and-cheek insinuations, and playful metaphors. Yet, the gaiety and sensuality observed inside these permissive and protective arenas should not be mistaken for (self-)indulgent enactments of alienating revelry or sexual impulse. Guided by the concept of serious play, samba dancers adapt or situate old concepts into new configurations inside the dancing circle as they alternate between (performative) desire and (physical) control or vice versa. Serious play functions, in particular, as a way of inscribing the past into the present as well as a way of fictionalizing "tradition" with ludic and pleasurable interactions. Also, samba dancers employ serious play in order to construct open-ended dialogical acts (*paidia*),[15] whose content is not invested in one specific meaning or goal. Rather, samba dancers insinuate sensual vacillations, pleasurable traces, and ephemeral erasures, whose improvised repertoire inevitably exceeds denotation.

Therefore, the ginga aesthetic equips thinking-moving bodies with the ability to articulate oscillating dialogues in which affirmations and negations are constantly being blurred into ambiguous (or contradictory) transitions. In other words, skillful samba dancers reproduce playful interactions whose directions and intentions are continuously being dismantled and re-routed, re-arranging the "nature" of their interaction. In doing so, their transient viewpoints are continuously re-arranged and renegotiated to instantiate sync-ness, apartness, or dissonance. Meanwhile, their polycentric and polyrhythmic bodies generate and dissipate ideas, all of which may assume abstract, symbolic, decorative, satirical, sensual, or even erotic tones. Ultimately, combining sensuality with playfulness, dancers assert themselves and negotiate power relations within their own encircling communities.

As a discursive metaphor, this "juggling" strategy (i.e. *jogo de cintura*) articulated between two dancers at the center becomes a forceful mode of action for a variety of symmetric and asymmetric negotiations, especially when performed between dancers of the opposite sex (or dancers performing opposing gender roles). The recognition of this rhetorical reversibility has led me to imagine this pliable game of seduction and control improvised between two dancers at the center as two sides of a coin: the negotiation of power/knowledge/desire over the *visible* space that the community circumscribes/recognizes. Combined, the structure of these protective and permissive events (circular formation), the system of bodily organization and knowledge production they employ (the ginga aesthetic), and the ambivalent collaborations they articulate (samba as process) inform the production of oscillating interactions (rhetorical ambivalence) where the ontological nature of self-othering monologues are constantly being put in "checkmate." In doing so, samba dancers are able to improvise new sequences in response to ever-changing contexts and, subsequently, to organize their personalized repertoire of syncopated improvisations into flexible choreographies of self-identification.

When dancers enact the role of a (heterosexual) couple in particular, the gendered division of labor is made visible in the interaction between the "female" and the "male" role-playing dancers. Yet, contrary to ballroom forms of samba dancing, where men lead and women follow, inside samba circles they both alternate leading and following one another, mostly without physical contact. Inside casual samba circles (e.g. at a backyard party or a street fair), both male and female dancers delineate strategies that range from mockery to seduction or indifference. Each and every action, gesture, look, or smile may also add new layers of gendered identification to what is being articulated, tinting actions with malicious, erotic, or playful flair. Moreover, based on their technical skills, psychosomatic confidence, and sexual disinhibition, both men and women may bargain equivalent amounts of time, space, and visibility inside the circle. As she drags her sandals across the floor, a dancer performing femininity may privilege movements of hip-syncopation that evoke active courtship, passive leadership, or vice versa. With both technical precision and sexual derision, the dancer's hips may hypnotize the audience with syncopated patterns, while her facial expression remains serene and detached. This high-affect juxtaposition may force her partner to chase or dance around her, begging for attention in return. Conversely, gender relations may take a different turn once a pair commits to move independently around each other in close proximity with their eyes locked, yet little physical contact, like a "gyroscopic glide" (Chasteen, 2004, p. 32). In such instances, the line between (masculine) assertion and (feminine) seduction tends to blur or shift positions with the same pliability that dancers wobble their hips in a figure of eight. With a sense of coolness, they negotiate, there and then, how to construct desire and when to make the pleasure they experience under their clothes visible to the public eye.

Beyond their processes of gender identification, the push and pull between seductive virtuosity and expressiveness, and technical control and self-assertion played out across these non-verbal compositions is contingent upon the dancers' situated knowledge and their relation to one another, as well as the context of their performances. Adults, for example, act very differently from children, and friends often dance closer and with more permissibility towards one another than strangers meeting at the center for the first time. Physical attraction and dexterity may also play important roles in the intensity with which dancing couples articulate either fierce seduction or playful harassment. In the last instance, dancers may access particular choreographies of gender and sexuality, based on the dancers' level of acquaintance with one another, their shared historicity, etc. in an attempt to have fun and yet remain memorable in the eyes of a particular group.

As my analysis above has sought to demonstrate, the beauty of samba dancing lies in the dancers' ability to both expose and hide intentions at will, providing open-ended metaphors to sparkle one's imagination. The implied ambiguity or indefiniteness cultivated within these practiced places conjures up a pliable mode of identification that, contrary to the set choreography of Globeleza, enables dancers to convey complex discourses of seduction and assertion, while avoiding being pinpointed into the passive position of an object of sexual desire. In contrast, Globeleza's set score, her exposed body, and the camera's colonial/patriarchal gaze leave little to the imagination. Upon taking a closer look, one also notices that, despite Globeleza's exquisite skills as a dancer, the use of high heels actually inhibits her ability to flex, stomp, or drag her feet and further oscillate her body in complex ways, as dancers with flat shoes do. In addition, her dancing sustains an elongated posture that clearly highlights her flat stomach, yet leaves little wiggle room to re-arrange the torso into polycentric and serpentine pathways. Hence, despite her eternal smile and exquisite sense of coolness, the upward linearity of Globeleza's symmetry leads her body into unstable and stiff positions. Since her feet can only attempt timid skips, she waves her arms contrary to her hips' oscillation in order not to lose her balance. At the end, her phallic dancing body seems to fight against gravity instead of embracing the offbeat "get down" quality of movement and playing with it.

4
Investigating the Articulation of Ginga in Capoeira Angola

From 1822 to 1825, the German painter Johann Moritz Rugendas (1808–58) embarked on a scientific expedition across Brazil. His voyage, carried out immediately after Brazil's proclamation of independence from Portugal, resulted in the publication of one of the most comprehensive portrayals of nineteenth-century daily life of that former colony. Containing over 100 images, Rugendas' ethnography houses the earliest known visual record of a capoeira game in Brazil. Briefly, the lithograph *Playing Capoeira or the Dance of War* (1835) depicts a lively gathering of black men and women watching a central wrestling match on the streets of Rio de Janeiro (Figure 4.1).[1]

I am instantly drawn to the fighter on the left. His tense musculature and his piercing stare, as well as his upper-body projection, suggest that he is about to strike an aggressive attack. Confirming my expectations, Rugendas gives the following description:

> Much more violent is another war game of the Negroes, jogar capoeira, which consists of trying to knock one another down with *head-butts in the chest,* which one dodges with *skillful side jumps and parrying*. While they are throwing themselves against one another, more or less *like rams,* sometimes heads run terribly into each other. Thus not infrequently the *prank turns into real fight* and a bloody head or a blade puts an end to the game.[2] (1835, p. 26, author's translation, emphasis added)

Similar to other accounts by European artists who participated in traveling expeditions across Brazil during the same period,[3] Rugendas' characterization of this "war game of the Negroes" as an animalistic ritual that can move from a prank to a lethal fight paints capoeira as a bizarre, or perhaps irrational, diversion. At the same time, his perceptive reference to "dodges with skillful side jumps and parrying" seems rather evocative of the languid, sinuous, and offbeat way of swaying the body that makes capoeira so fascinating to look at. In fact, upon closer examination, I realize that the man on the right seems to be dancing with rather than fighting against his opponent. With a nonchalant expression, he tilts his upper body towards his front leg, while sliding his hips backward. Maintaining a dynamic balance, he sweeps both arms across his bare chest (from right to left) and back, as if rowing, while scooping up his back (left) foot off the ground. Is it possible that these playful combats once functioned as rehearsals for slave revolts? Or was this war dance simply a kind of "docilized" act of revelry, a necessary escape valve that prevented more serious uprisings?

I look around the lithograph and I am further puzzled by Rugendas' inclusion of a drummer and a few celebratory gestures in this carefully constructed scene, as well as his omission of chains and other pieces of apparatus indicating the exercise of colonial power over these enslaved bodies (a condition symbolized by their bare feet). Were these

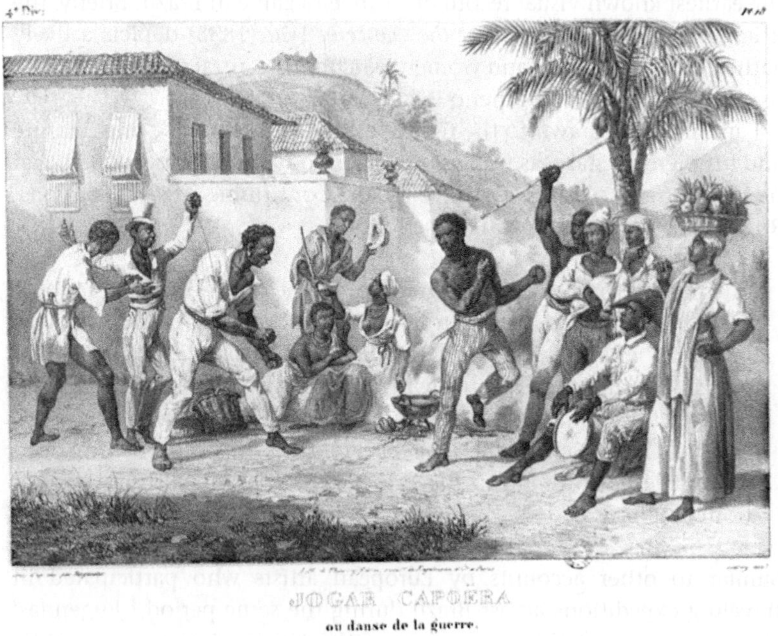

Figure 4.1 Johann Moritz Rugendas' *Jogar Capoera ou Dance de la Guerre*, 1835

internationally displaced and exploited captives as happy as Rugendas depicted them or is this archival document, as dance scholar Marta Savigliano puts it, "the seemingly harmless side of exploitation, cloaked as it is in playfulness and delirium; a legitimate practice of discrimination, where the otherwise secretive fantasies can be shared out aloud" (1995, p. 169)? Despite his discerning eye, Rugendas' colonial gaze contributes to the exotification of this ethno-cultural activity.

In an attempt to revert the exoticism with which this martial art has been archived since colonial times, in this chapter I examine the underlying mechanisms through which capoeira players weave together dancing, fighting, and playing *efforts*[4] into call-and-response improvisations.[5] In order to accomplish an in-depth analysis of the relevance of ginga aesthetic within the practice of capoeira, I focus on one particular lineage known as *capoeira angola*, whose genealogy may be traced back to Mestre Vicente Ferreira Pastinha (1889–1981). More importantly, my study departs from the understanding of capoeira angola as an embodied way of articulating ideas corporeally (process), rather than a formalized technique with codified rules or a game with well-defined goals (product). Therefore, although critical and detail-oriented, my analytical investigation is not committed to constructing a taxonomic catalog of all possible maneuvers available in capoeira angola. I will not attempt to list, for instance, a codified lexicon of movements deployed by capoeira angola practitioners, taking into consideration that their movement vocabularies, and the names associated with them, tend to vary from region to region and from time to time, based on internal and external influences. Instead, I offer a set of guiding tools that make it possible to unpack the underlying logic of bodily organization informing both the utterances improvised between any two players (based on rehearsed and improvised vocabulary) and these practitioners' judgment of taste (Bourdieu, 1984), including the embodied values and attributes they assign to abstract concepts, such as beauty, grace, elegance, and pride. Hence, rather than establishing a fixed taxonomy or a codified manual, this analytical "tool kit" provides a pathway to understanding how the ginga aesthetic shapes up the thinking procedures tying together this improvisational, dialogic, and process-oriented practice as a whole. Finally, my examination identifies and critically analyzes the means through which practitioners of capoeira angola deploy the Africanist movement system identified in this book as the ginga aesthetic to recuperate-cum-invent a diverging way of thinking about and moving across the world (i.e. epistemology) whose scope exceeds or differs from that in Western thought. In doing so, this chapter seeks to shed light on the following question: how do these improvised body games of call-and-response centered on the ginga aesthetic generate or articulate choreographies of (self-)identification?

Preliminary considerations

Before I delve into the "nuts and bolts" of capoeira angola and its underlying system of bodily organization and knowledge production, I think it important to situate the trajectory of Vicente Ferreira Pastinha, or Mestre Pastinha (1889–1981), a mulatto born to a Spaniard father and a black mother and considered by many to be the "founding father" of capoeira angola. Though born at a time when capoeira was considered an illegal practice, Pastinha learned this martial art from a freed slave named Benedito and continued to practice it during his teenage years while undertaking military service in the Marines. In around 1912, he gave up capoeira due to police persecution, but later resumed his practice, attending following the encouragement of friends in the late 1930s. In 1941, the now renowned Mestre Pastinha was invited to spearhead a movement against Bimba's "modernization" of capoeira and, after a few attempts to establish a formal school, in 1949 he opened his *Centro Esportivo de Capoeira Angola* (Capoeira Angola Sport Center), which was formally registered in 1952.[6]

Over the years, Mestre Pastinha's school gained wider recognition and he became a celebrated public figure amongst the intellectuals of Bahia. By the 1960s, his capoeira angola school and especially his Sunday rodas were attended by a mixture of local capoeira mestres (literally "master teacher"), students, tourists, and intellectuals, such as the Argentinean illustrator Carybé, the French photographer/anthropologist Pierre Verge, the novelist Jorge Amado, and the sculptor Mário Cravo, all of whom contributed towards consecrating Pastinha as an icon of Afro-Brazilian culture. Around that time, Pastinha met Neves e Sousa, an Angolan artist who affirmed that capoeira resembled the *N'golo*, or "zebra dance," a male rite of passage traditionally practiced in Angola, thus reaffirming the connection between the Afro-Brazilian martial art and the African continent. Considered by many to be the "philosopher of capoeira," Mestre Pastinha published a book and recorded an LP during his lifetime, and left a few other illustrated manuscripts, among which it is worth mentioning his *Metafísica e Prática da Capoeira* (*Metaphysics and Practice of Capoeira*). Some of Pastinha's famous aphorisms, spread via oral tradition, include "capoeira is everything that the mouth eats," which connects capoeira to the mythology of Exú, the trickster deity of Candomblé, also known as the messenger between the world of the living and the world of spirits, and "the body is a great system of reason," a statement that finds an intriguing resonance with Nietzsche's thought.

Upon his death, Pastinha's legacy continued to be exercised by his disciples, especially Mestre João Pequeno (João Pereira dos Santos, 1917–2011), who continued to teach in Salvador until his death at the age of 93, and Mestre João Grande (João Oliveira dos Santos, b. 1933), who currently lives and teaches in New York City. Most significantly, with the re-democratization of the country in the 1980s and the subsequent rebirth of black political movements of militancy against racial discrimination (i.e. the *Movimento Negro*), socio-cultural manifestations centered on Afro-Brazilian heritage and identity such as capoeira groups experienced a "re-birth." Like Ilê Ayiê, Olodum, and other *blocos afros* (Afro-Brazilian krewes), modern capoeira angola groups such as the historical Grupo Capoeira Angola Pelourinho, the Center of Capoeira Angola of Mestre João Grande, the International Foundation of Capoeira Angola, the Grupo Nzinga de Capoeira Angola, the ACANNE, the Grupo de Capoeira Angola Zimba, and the Grupo de Capoeira Filhos de Angola, to name but a few, have contributed to situating capoeira angola's movements, rhythms, lyrics, instruments, symbols, tales, and actions in relation to the African presence in Brazil and their desire to (re)shape their inter-subjectivity beyond coloniality's matrix of exploitation. Collectively, these distinct organizations have worked towards rescuing aesthetic and philosophical knowledges historically persecuted and marginalized, for they represented blackness.

Though the origin of capoeira continues to be the subject of heated debate, its historiography also indicates that this transnational martial art has been subjected to a variety of gradual transformations over time. As outlined in the first part of this book, capoeira has incorporated new concepts and forms over time, and it has equally generated particular meanings and discourses in different scenarios where it flourished, especially in port cities, due to the circulation of distinct peoples and ideas across particular points in time. Therefore, it is safe to assume that capoeira's repertoire has absorbed elements converging from a variety of African traditions, along with (or in spite of) other transcultural intersections and "contaminations" (see the Introduction), possibly including different Amerindian, European, Middle Eastern, Asian, and later North American influences. Like hip hop and other embodied practices of resistance circulating throughout the African diaspora, capoeira nurtures a distinctive mode of being-in-the-world that, to borrow Diana

Taylor's understanding of performance, "carries the possibility of challenge, even self-challenge, within it" (2003, p. 16).

Since the 1990s, capoeira has been riding the wave of globalization, so to speak, disseminating its seeds worldwide with an accelerating velocity. Although historically practiced almost exclusively by black and mulatto men from Brazil, a growing number of women and men living in places such as Tel Aviv, Tokyo, Mozambique, New York, and Paris regularly practice this playful war dance. Despite their gender, ethnicity, race, social class, cultural background, or place of birth, through the practice of capoeira, many have also incorporated the ginga aesthetic into the construction of their own processes of self-making. In doing so, these international practitioners have been instrumental in the valorization and proliferation of this non-hegemonic way of sensing the world and acting upon it.

Until very recently, events and activities relating to capoeira were often associated with roguery or idleness rather than profitable work. However, in this chapter I would like to propose that we investigate what this practice accomplishes or produces on its own terms, instead of framing capoeira

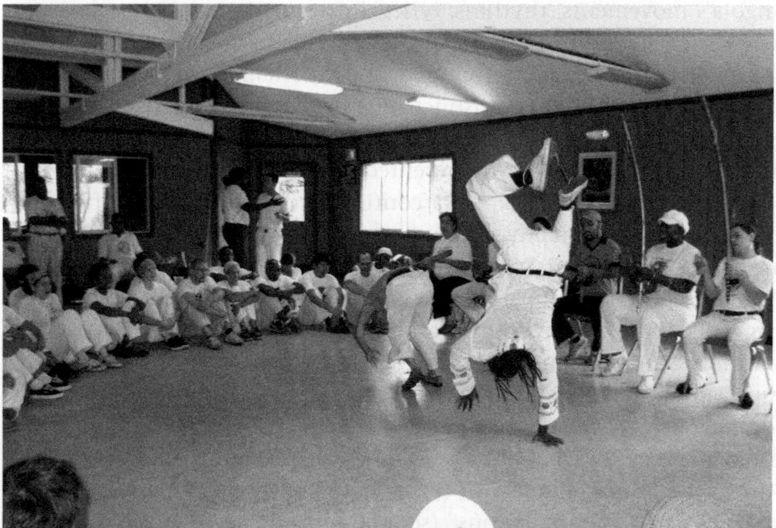

Figure 4.2 Mestre Cobra Mansa playing capoeira in Chicago © Cristina Rosa, 2008

in opposition to efficiency, in the (capitalist) sense that this term has acquired in modern times. Briefly, a capoeira circle (*roda de capoeira*) is an Afro-Brazilian performative practice that includes elements associated with dance, martial arts, music,[7] theatre, and oral history. Like *rodas de samba*, capoeira circles function as a social/ritualistic event where participants are invited to watch, entertain, and interact with one another at all times. For example, the structural organization of capoeira circles permits participants to rotate between the position of central players, leading singers, musicians, and audience according to their level of skill (e.g. their ability to play a particular instrument or improvise call-and-response songs). In addition, the call-and-response musical interactions of live percussive orchestra,[8] which take place alongside the perimeter of the human circle, inform – and are formed – by the non-verbal dialogues improvised between two players at the center, that is, the capoeira game (*jogo de capoeira*).

Despite the communal atmosphere and interactive mode of action, the center and the rim of the roda constitute two distinct realms, each organized according to a different set of guiding principles and operating towards a distinct set of purposes. The rim or perimeter of the circle may be understood as a set and hierarchic structure. While participants may accumulate functions and rotate between positions, not every place has the same value. Within the orchestra, there is an additional system of stratification, based on the level of importance associated with each instrument; hence, some positions carry more value than others. The center of the percussive orchestra – a notable position – features three musical bows known as *berimbau*.[9] Both a musical instrument and a weapon, the berimbau symbolically stands for African musical and oral heritage as well as the struggles of Afro-Brazilians. Each musical bow is equipped with a resonating gourd of a different size, which is thus tuned to a different key. The berimbau with the largest gourd and deepest sound, known as *gunga*,[10] marks the highest position of power. Like an epic troubadour or griot, the person playing the gunga commands the musical aspect of the performance and controls the mood of the event, displaying his or her ability to execute both standard riffs (*toques*) and virtuoso improvisations (*dobradas*), dictating the pace of the orchestra, and guiding the other instruments.[11] In addition, "the gunga" also oversees all the actions alongside and within the circle, calling songs, controlling the game enacted at the center, and overall holding together of the structure of the roda.

The center of the circle is a completely different arena. Unlike the perimeter, all participants are welcome to step into the circle to play a combat game regardless of their level of skill or seniority/prestige in that group. In capoeira angola, the games enacted at the center adhere to the following sequence of events. After practitioners gather and form a circle, two players enter the circle and squat down at the "foot" of the berimbau (in reverence). Next, the person leading the circle commences the ritualized performance with a praise song known as *ladainha*, after which the two players are given a sign to begin playing. In most cases, the person commanding the ritual also stipulates the duration of each game, although players may choose to "shake hands" and end their section. At the end of each game, players congratulate one another amicably and step off the circle, using the same "entrance" point. Depending on the nature of the event, the number of participants, and the skill level or fitness of each player, these jam sections may endure from one to six hours, with each game lasting anywhere from five minutes to half an hour.

Beyond its atmosphere of sociability, the improvised dialogues between two players at the center, as well as the musical call-and-response interactions between the center and periphery, promote both tradition and innovation. As Greg Downey argues, for instance: "The variability inherent in capoeira training guarantees that the art itself is no single thing. Rather, students reinvent capoeira in their own idiosyncratic fashions, even under the watchful eyes and steady hand of a mestre" (2005, p. 39). This is a vital argument against the discourses that advocate for the "traditional" capoeira, whether in its Brazilian or African form. By acknowledging transformation and creativity, the aesthetic and philosophical fundaments of capoeira do not become something players must wear like a vest or armor, but rather a set of guiding principles with which to articulate pre-established and innovative statements.

Articulating knowledges otherwise

At first glance, each capoeira game may be understood as a match of honor: a strategic battle to territorialize the field of action and stand apart, while connecting to the circumscribing group. In other words, while the corporeal games enacted within circles of capoeira angola proudly promote the group's pride towards Afro-Brazilian heritage in general, it also enables individuals to figure out personalized pathways to articulate their own process of self-fashioning and thus gain the respect of their peers. In that sense, these jam sessions function as ethno-cultural events where community members are given a chance to "show off" their (individualized) physical abilities, mental capacities, and overall creativity, while bringing pride towards their resisting processes of (collective) identification. In addition, upon closer examination, it becomes clear that across their call-and-response dialogues, capoeira players propose metaphorical questions to "test" the other in the hope that their opponent will eventually run out of answers or will be unable to improvise them in a timely fashion, thus disrupting their sense of coolness. Rather than games with a final goal (*ludus*), within these performative arenas, the two angoleiros at the center co-choreograph open-ended and playful interactions (*paidia*) of self-othering identification that push and pull one another's physical and mental capabilities. When fully realized, these non-verbal dialogues articulate asymmetrical negotiations of power relations where aggression is intertwined with other elements such as derision and seduction. Instead of producing truth and meaning, in the Western (Platonic) sense of the term, these exercises of agility, dexterity, strategic planning, and theatricality amount to a dynamic, pliable, and decentralized way of articulating ideas, informed by concepts such as coolness and serious play. There is, I repeat, no "end goal" or "finishing line." As further discussed below, capoeira angola players deploy the ginga aesthetic in an attempt to remain "in the middle": neither too hot, violent, and inquisitive nor too cold, passive, and reactive. I will refer to these thinking-moving dialogues the process of capoeira angola.

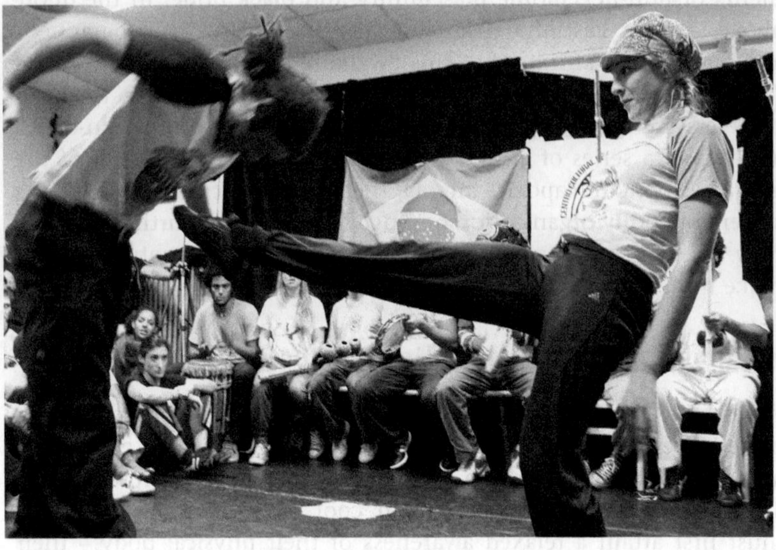

Figure 4.3 Contramestre Fubuia (Itaparica, Brazil) and Professor Chiclete (Lille, France) playing capoeira in Paris © Cristina Rosa, 2012

The function of ginga within capoeira

On the LP that Mestre Pastinha recorded in the 1960s, the father of capoeira angola offers an overview of the idiosyncratic maneuvers performed both inside and outside the roda. Regarding the swaggering way of moving, or gingado, characteristic of "old school" capoeira players (*capoeiristas*), Pastinha points out that:

> In my time, when I was a capoeirista ... There were capoeiristas who *walked around twisted*, but twisted in a way that *nature did not make him* ... The hat thrown to one side ... And he walked in the middle of the street with that sway (*gingado*)![12] (Assunção, 2005, p. 111, emphasis added)

As Pastinha indicates in the quote above, during the first part of the twentieth century, many distinguished players – or troublemakers – defied common decorum as they walked the streets with languid yet disturbing movements of bodily syncopation. Perfected through the practice of capoeira, their unique "swayed walk" was employed as a signature way of interacting with the world, a choreography of self-identification that tied performative defiance to collective enactments of non-hegemonic masculinity and black pride. In the quote above, Mestre Pastinha insinuates the idea of a "choreographed" action when he affirms that this "walking around twisted" was not "natural." In other words, it was not the result of a defect in the locomotive system of the person walking. Below, I will offer an in-depth analysis of a series of efforts associated with the deployment of the ginga aesthetic in capoeira angola, (i.e. the specifics of the capoeira's system of bodily organization or gingado) in order to further examine how a person might go about moving "in a way that nature did not make him."

Maintaining an alert relaxation: coolness

Kinesthetically speaking, the ginga in capoeira angola physically materializes the dynamic balance of opposing forces commonly associated with West African concepts of coolness (Thompson, 1966; Tavares, 1984). In order to remain cool while in motion, players must first attain a relaxed awareness of their physical body – their motor-sensory system, their breathing, their facial expressions, etc. Angoleiros sway parts of their body with a tension-free attitude, maintaining a continuous flow across propellers (feet), pairs of scales (hips and shoulders), the vertical flexible axis, and the limbs. Muscle tensioning or the hyper-extension of joints, for example, hinders the flow of back-and-forth interactions. In addition, all five senses work in tandem to gain a serene and multi-layered awareness of the surroundings so that players can pay attention to what is happening at the center as well as the periphery, in front and behind them. Controlled breathing, in particular, may guide the beginning player into this focused-yet-detached mode of action, which is sometimes referred to as "the capoeira's trance" (see Decânio Filho, 1997). The emphasis on this rehearsed mode of attention, namely a "restful alertness," pushes players to act and react promptly, either maintaining a gliding and seamless pliability and a continuous flow or composing rhythmic juxtapositions and dissonances (e.g. discrepancies in time) as they shift gears. Metaphorically, their swaggering coolness may be compared to the corporeal sensation of standing on a sail raft or small riverboat, wading through water.

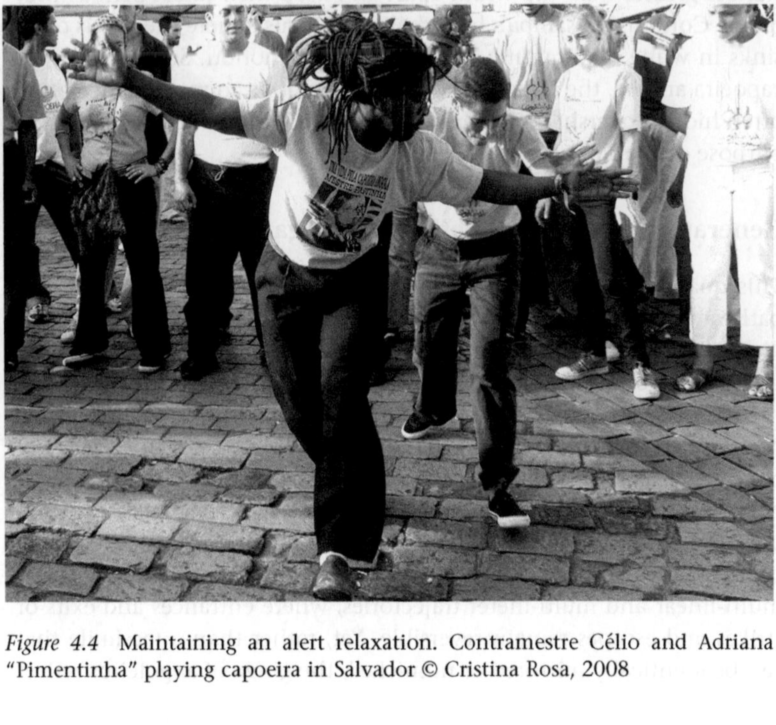

Figure 4.4 Maintaining an alert relaxation. Contramestre Célio and Adriana "Pimentinha" playing capoeira in Salvador © Cristina Rosa, 2008

Fragmenting the body: visual dissonance

This effort refers to the predominance of articulations between the ground and fragmented parts of a polycentric body as the player tilts the hips sideways, slides the shoulders, or thrusts the two pairs of scales, generating high-affect juxtapositions. These "breaking" dialogues constitute a triggering event, from which other pathways of movements, gestures, and intentions are built. The offbeat shifts and turns emerge from the internal awareness of one's decentralized corporeality. Subsequently, each player may reproduce a wide range of syncopated patterns of movement, sliding one leg towards the back, leaning the chest forward and back, swinging both arms outwards then pulling to one side like rowing, or propelling one or both feet upwards into a kick. In sum, it prepares one to, as Mestre Pastinha suggests above, walk "around twisted, but twisted in a way that nature did not make him." At the same time, the apartness with which players fragment their bodies during oscillating interactions may also be triggered by external riffs, such as the circular dribbles of an opposing player or the tonal polyphony of the live percussive orchestra. Overall, angoleiros privilege stances greater than hip-width apart and keep their knees bent as they move. Contrary to samba's hyperbolic hip thrusts, here the pelvis often sinks in while maintaining its wide range of motion. Subsequently, in capoeira angola, the upper body's sinuous actions counterbalance – and often hide or overshadow rather than expose – the direction, timing, or purpose of the hip-play.

Generating syncopated pathways: ginga or gingado

This quality of movement refers to the predominance of multi-linear pathways of oscillation (swayed action) resulting from polyrhythmic dialogues between bodily parts, especially the hips and feet. As players dislocate across time and space, often coming in and out of sync with the main riff, they step from side to side, back and forth, or combine both directions (triangular oscillation). In the last case, the surface of contact with the ground (e.g. the soles of the feet) outlines triangular patterns of dislocation, such as side-back-side or side-front-side. By combining side-to-side and back-and-forth oscillations into a multi-angular transfer of weight in space, angoleiros employ the gingado to construct multi-linear and multi-meter trajectories, where entrances and exits or strikes and escapes remain reversible. Yet, rather than a facsimile that can be identically mirrored or mimicked, the ginga of angoleiros reflect their personalized style. It informs, first and foremost, the players' unique ways of appropriating aesthetic principles and standing apart rather than reproducing a standardized or idealized uniformity. Some players choreograph a minimalist gingado, for example, where the arms stay mostly bent and near their upper torso, relaxed but alert, and the hips wiggles just enough that they can shift directions seamlessly. Others tend to flex and curl their entire bodies, running their fingers across the ground as they lean forward. A few are able to twist and swerve as though their torsos were made of rubber, without losing sight of their opponents. Others wobble and stagger as if they were drunk or had a crooked leg. Like a *(re)cycling signature*, each player reiterates through his or her swagger an individualized way to follow rhythm, connect to others, stand apart, negotiate space, hide intentions, and avoid direct confrontation. In doing so, the players articulate their bodies in creative (or unpredictable) ways. Without talking, the players at the center of the circle generate playful dialogues of perception, intuition, and imagination with one another.

Extending ginga across the body: serpentine pathways

This quality of movement refers to the predominance of spiral and wave-like motions – beyond the vertical interaction between the hips and feet – in order to generate soft-spoken rippling effects across the entire body. The initial syncopation between the hips and feet, for example, may be extended towards the torso, arms, and legs, triggering back-and-forth oscillations between the center and periphery. These snake-like actions push players to transfer weight across decentralized body parts in space as well as to wave their flexible spines to evade an attack with a detached elegance and efficient subtlety. Following the "alert relaxation" concept discussed above, the upper limbs often undulate rhythmically in space in response to the initial offbeat swing of the core. In the particular case of the triangular oscillation, or basic gingado, the syncopated sway of the hips is often accompanied by a radial or tangential swing of the arms around the torso. Although never hyper-extended or held at rigid angular positions, the arms both protect and expand one's kinesphere with a rowing-like pattern of motion. In this way, although the hips continue to trigger juxtaposed and dissonant transfers of weight across time and space, the expansive articulation of limbs outwards overshadows the visibility of their range of motion. Besides, these rippling effects, especially in conjunction with peripheral areas of the body such as arms and legs, corroborate to "hide" or

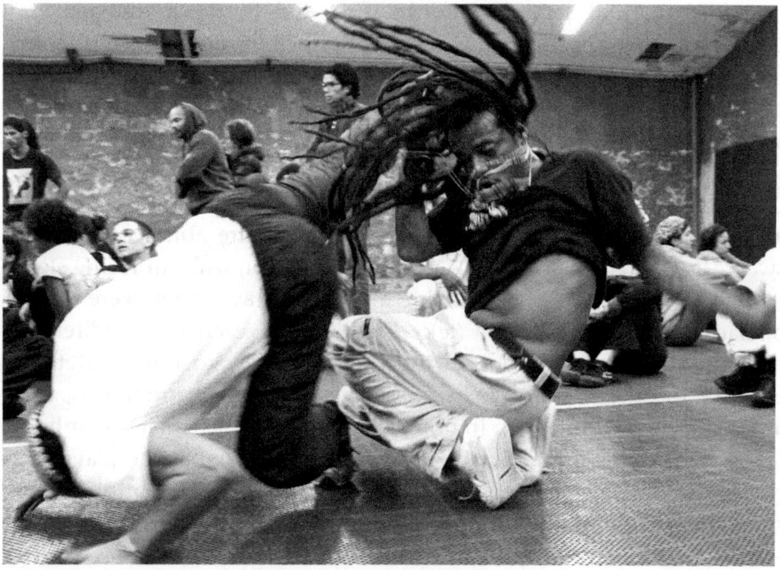

Figure 4.5 Serpentine pathways. Professor Xixarro playing capoeira in Paris © Cristina Rosa, 2012

"masquerade" the centralized articulation of the hips – in other words, direction or intention.

Establishing the equivalence of five points of support: radial symmetry

This quality of movement refers to the use of different peripheral parts of the body to support weight, as the center of gravity transitions from one side to another. Instead of a tree-like structure, angoleiros articulate their bodies as radial or rhizome-like structures. In capoeira angola, the five major points of contact with the ground are the top of the head, the flat open palm of both hands, and the soles of both feet. All five points of support are, or should be, treated as equivalent.[13] Through variations on the basic pathway of oscillation, players circle back and forth in space, transferring their weight from one (or more) of those five points of contact to another. As players adjust and reposition themselves across space, the hierarchy of their own bodies is often inverted or downplayed. In addition, it is not uncommon to observe an angoleiro with three or more points of support in direct contact with the ground, as he or she spirals across the shared space.

Floating through the ground

This quality of movement refers to the familiar interaction players establish with the ground as they sustain a rhythmic flow and construct pathways of movement. In addition to the swing of the hip joint, players generate momentum by either transferring weight towards the ground (launching/diving/falling) or moving away from it (pushing/jumping) with any of the five points of support. In instances when a player moves horizontally across the floor, the upward swing of the hips may be employed to get a player off the ground. This is especially useful in cases when a player is swept off his or her feet or loses his or her balance quickly. In addition to providing players with a wider sense of stability and security, on the ground maneuvers are often employed as a "safe" way to penetrate the other's personal sphere from the ground up.

Wheeling the vertical axis

This quality of movement refers to the predominance of patterns of movement where the pelvic area functions as a wheeling device, propelled by the push-and-pull interactions with the ground. In practical terms, this means that in addition to moving in seesaw and/or triangular pathways, a player may swing his or her hips on their axis (like a spinning wheel), enabling both superior and inferior limbs (plus the head) to touch or propel from the ground. Angoleiros employ their wheeling hips to transfer the support of their bodily weight across their primary points of support described above, while projecting their center of gravity outwards in space or downwards, toward the ground. Following centrifugal pathways from the pelvic area outwards, the overall movement of the body may acquire either an undulating performance of uninterrupted flow or a high-juxtaposed interaction that disrupts the (expected) fluidity with movements of contrariety. In the "au," or cartwheel move, for example, (one or both) hands (and the head) may touch the ground in support of bodily weight, while waving the legs in the air. Besides, in the case where the hands occupy the position of propellers/supporters of bodily weight, the legs and feet may be extended into a top-down attack, used as a pre-eminent protection (e.g. knees bent near the stomach), or otherwise be "freed up" to gesticulate ideas (e.g. aggression or mockery; see the discussion below). In doing so, they contribute towards, among other things, the territorialization of the performative space and negotiations to control the game (or its range of possibilities).

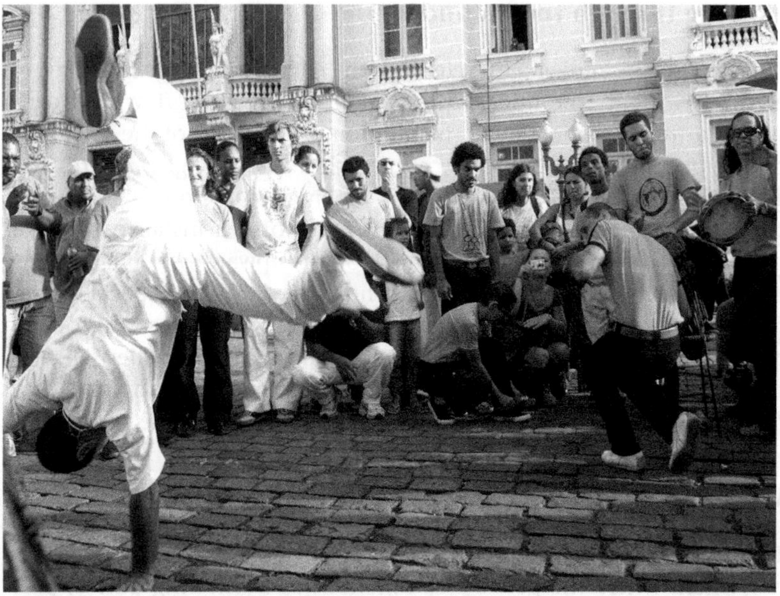

Figure 4.6 Wheeling the vertical axis. Mestre Poloca and Márcio playing capoeira in Salvador © Cristina Rosa, 2008

Hanging at the edge of balance: finding a dynamic equilibrium

This quality of movement refers to the use of different parts of the body (the limbs, the torso, the head), which are not (momentarily) in contact with the ground, to create, maintain, or restore balance during weight transfer. If the player launches his or her upper body sideways towards the ground, for example, the left arm may draw an arch from side to top, counteracting the force of gravity. In the particular case of headstands and handstands, the muscles and joints of the pelvic area assume a new function. Instead of oscillating back and forth, in this case the pelvic area's muscles and joints further counteract the force of gravity, controlling the dynamic equilibrium of the flexible spine from the top down. Furthermore, in the case of a "queda de rim" (literally "descend on the kidney") – a modified handstand where the player moves towards the ground and folds one of his or her arms under the kidney region – the legs may wave in the air like scissors, preventing the player from losing his or her balance and collapsing on the ground.

At other times, a player may bend the legs near the chest, curling like a ball, while keeping the opponent at a safe distance.

Lurching off-balance: expanding the equilibrium dynamic

This quality of movement refers to the predominance of movements generated from the displacement of the center of gravity towards (or beyond) the verge of disequilibrium. In capoeira angola, static and/or stiff positions are seldom sustained. Since players are constantly rocking their weight from side to side, they are continuously in the process of losing and re-establishing balance. This back-and-forth dynamic of (un)balance defies both the body's (static) footing and functionality. Always pushing and pulling the body's center of gravity toward its threshold (or at least faking it), skillful players combine flexibility, strength, and acrobatics to produce uncanny, albeit delightful or ludicrous, pathways of movement. Walking on one's hands is a good example of this dynamic displacement: as a player swings his or her hips and legs up in the air, the slight projection of the center of gravity provokes the whole body to dislocate or precipitate even further. The body tilts as a consequence of this (un)controlled balance. The hands move to re-adjust or "catch" the newly established center of gravity. Following the alert relaxation awareness, joints such as the knees, elbows, and the torso are kept slightly bent, as they function as shock absorbers that produce a seamless transfer of weight.

The role of ginga within capoeira

In *Capoeira Angola* (1964), Mestre Pastinha confirms my expectations regarding the roles that the ginga aesthetic assumes in this martial art form. For the founding father of capoeira angola:

> The word "ginga," in capoeira, means a perfect coordination of bodily movements that the capoeira player executes with the goal to *distract the attention* of his adversary, in order to make the other vulnerable to the application of his attacks. The ginga movements are suave and very flexible – which confuses the ones not familiarized with capoeira, making them an easy prey for an aggressor that is familiar with this modality of fight. Within the ginga one finds the extraordinary *malícia* of capoeira that stands as its fundamental characteristic. The ginga in capoeira also has the great merit to develop the equilibrium of the body, lending itself to *the softness and grace proper of a ballet dancer*. (1964, p. 50, emphasis added)

As Mestre Pastinha outlines here, in addition to helping players to develop a graceful and thus non-threatening way of moving, the ginga aesthetic has a strong connection to the concept of *malícia* (meaning cunningness or astuteness) through which players both attract, distract, and even attack their adversaries.[14] Based on Mestre Pastinha's observations, below I discuss three concrete ways that this particular system of bodily organization informs the ideas that players articulate through movement.

Departing from the ginga aesthetic, angoleiros employ a wide range of fight-like efforts in order to choreograph prowess, aggression, and fortitude by throwing their arms, legs, head, and pelvis around. Movements such as head butts and kicks amount to *tactics of aggression.* In particular, the predominance of maneuvers instantiated from the hips, which project the legs and feet outwards towards the opponent (e.g. kicks and sweeps) in order to attack or otherwise intimidate the opponent, results in a "virilization" of ginga. Keeping their feet wider than hip width, for example, players rely on the swaying hips to perform a wide range of firm decisions, impingent attacks, courageous moves, and expansive claims to space, all of which denote virility and aggression. In this case, the gingado evokes physical force and violence. Quite often, these maneuvers demand strength, courage, and control, qualities commonly associated with masculinity. Practically all boots are designed to dislocate the opponent and (possibly) "mark" him or her, or at least threaten him or her with such a possibility.[15] Furthermore, the angular and forceful projection of the pelvis forward, indicating but not executing a kick, may be enough to cause the opponent to retreat. This virile display of strength, vigor, and muscularity has always been part of capoeira's repertoire. In its extreme versions, however, it may perform (or be perceived as) irrational or uncivilized brutality.

Contrary to the display of violence and virility discussed above, there are a number of dance-like efforts in which the gingado is implicated in *tactics of illusion* with which to achieve "the softness and grace proper of a ballet dancer," as Mestre Pastinha puts it. Angoleiros engage in embellished and eloquent drifts instantiated from the hips, for instance, which seek to deliberately dissimulate the direction of a kick or to mystify the timing of a sweep with an embodied sense of coolness. As expected, this kind of corporeal allurement functions as a way to negotiate power indirectly, while avoiding the use of physical force or other coercive appeals. The deployment of various hip-driven and serpentine-like qualities of movement is often employed to "hypnotize" the opposing player and the participant-audience.

Departing from hidden moves, soft maneuvers, low-key attitude, or sinuous articulations, these tactics of illusion lead to a delightful form. Constantly shifting the weight of their bodies with a languid and liquid sense of continuity, for instance, angoleiros often fluctuate across delicate stances, spinning on the tips of their toes, rolling between soft and rounded gestures, flexing their torsos or otherwise improvising movements that evoke weightlessness and vulnerability. Yet the asymmetrical sinuosity with which players glide back and forth in space to produce an artful game is seldom enacted in vain. Although senior players invest a considerable amount of energy towards moving with the slippery beauty or elegance of a snake, a chameleon, or a fish,[16] these tactics of illusion have the practical function of feinting directions, hiding intentions, or otherwise distracting the other. As Mestre Cobra Mansa defends, players demonstrate the efficiency of this tactic when executing maneuvers that are aesthetically pleasing, but also well executed and applied at a relevant moment. More often than not, movements of inductive seduction are followed by unpredictable, and hence efficient, counter-attacks.

In addition to aggression and illusion, the use of ginga to twist the body into preposterous and risible positions, such as playful inversions, awkward jumps, or uncanny postures, enables a player to articulate a series of persuasive efforts grounded in serious play. Above all, these play-like efforts or *tactics of humor* corroborate with their overall mental juggling and negotiating. In other words, the use of hip-play in conjunction with physical jokes and expressions, mockery, irony, parody, or even laughter help break the ice of their call-and-response movement battles. These clowning tactics enable angoleiros to relax and maintain an open attitude towards one another, often preventing a game from escalating into a violent assault. Playful sideways contortions and contractions, for instance, may tint a player's personalized way of moving with sadistic humor and quasi-innocent roguery. Likewise, angoleiros instantiate both trickery and derision when twisting and sliding their bodies like a drunkard or waving their legs in the air like the leaves of a banana tree. At other times, players may purposefully "show off" the malleability of their spinal column with hyperbolic twists and turns, strategically hiding their ulterior intentions or simply working out a style no one can deal with. Although this is often employed as a diversion, clowning or child-like attitudes give players a chance to stage their moving bodies in spectacular, yet perhaps wily, ways.

Illusion, aggression, and humor[17] may work in tandem with other ranges of tactics not discussed here. At any rate, once angoleiros learn to articulate each of these efforts, they are also advised to avoid privileging one to the detriment of the others, as it may lead them to acquire a fixed reputation. The predominant reiteration of fight-like efforts, for instance, may be perceived (or rejected) as hyper-aggressive behavior. Conversely, an overflow of dance-like or play-like efforts within one's personalized repertoire may be perceived as an indication of docile weakness or lack of objectivity. In that sense, when considering the efficacy of each tactic in a particular situation, one must also take into consideration the physicality of both players involved in the process. A wise player who is taller and visibly stronger than his or her playing partner may strategically choose to avoid the extensive use of expansive and virile maneuvers in that particular game, otherwise he or she could project a disproportionate self-image of irrational brutality. Instead, this player may choose to enact a less threatening choreography of self-identification that privilege dance-like or child-like efforts in order to increase his or her chances of disguising any combative intentions and, subsequently, may execute unpredictable and perhaps fatal attacks. By the same token, smaller or slender players may opt to prioritize the virile use of the hips and wider stands as a way of expanding their kinesphere and territorializing the shared performative space. In that scenario, short players may also take advantage of fight-like maneuvers as a way of balancing out or counteracting the frailty commonly associated with the size of their physical bodies. In sum, seasoned players embody and disembody each of these tactics, as well as others not discussed here, with a purpose in mind. In the context of a specific game, players jump across a wide range of imagistic utterances, both attracting and distracting the other. In capoeira angola, more specifically, the fake and flickering dialogues enacted between players at the center function as fractured/dissonant articulations and figurations. In doing so, they enact flexible choreographies of (self-)identification, whose embedded trickery, double-entendre, or polysemy always already differ and defer fixed identities.[18] This process becomes more evident in the call-and-response structure of their non-verbal discourses.

Call and response: choreographing improvisations within capoeira angola

As previously stated, angoleiros employ a ginga aesthetic to organize their own moving bodies and to construct dialogical and multi-linear trajectories as they move across time and space. Moreover, the ginga aesthetic plays a fundamental role in the construction of call-and-response interactions between players. Hence, despite the specificity of each context and the "apartness" of their personal styles, the ginga aesthetic structures both the form and content of their non-verbal conversations, whose syntax is oriented by concepts such as asymmetrical fluidity, reversibility, and polysemy. Below I will outline some of the most significant movement patterns over which this polycentric and polyrhythmic movement system, i.e. the ginga aesthetic, orients and shapes up their call-and-response improvisations.

Within the game of capoeira angola, eye contact precedes all physical interactions. Each game starts with an acknowledgement stare as players enter the circle and crouch at the "foot" of the berimbau (*ao pé do berimbau*).[19] Likewise, each game ends with a disengagement of sight after players return to the starting point and shake hands. As a rule of thumb, angoleiros seek to maintain a continuous bond through sight, even when not facing one another, in an attempt to read and interpret each other's body language, and perhaps second-guess the other's intentions. Besides, this visual engagement also influences the nature of their game. In general, players try to keep their eyes locked into each other's actions as much as possible instead of looking at the floor or where they are going. In a handstand, for example, seasoned angoleiros try to face forward instead of staring at their own hands.

Throughout the game, eye contact also enables players to synchronize their gingado kinesthetically. Despite the lack of extended physical contact, as players sway their bodies from side to side, they look for ways to watch each other's steps, often mirroring or at least gazing at one another. This mirroring strategy, conducted through sight, enables them to move with safety and predictability until they find an "open space" or "gap" through which to intervene. Furthermore, some angoleiros, by looking into their opponent's eyes, are able to read their hidden intentions. As players begin to spiral in and out of each other's personal sphere while inverting and reverting their vertical axis, keeping track of the opponent's position becomes one of the most challenging yet essential tasks to sustain across the game. Conversely, when a player loses sight of the other momentarily – a sign of vulnerability – the fluidity of the game breaks down and, more importantly, this may lead to unforeseen predicaments, such as a swift takedown or a treacherous "sweep" attack (*rasteira*).

Secondly, capoeira players sense and synchronize their call-and-response interactions through auditory perception (hearing). In capoeira angola, all movements, gestures, and intentions are (or should be) rhythmically coordinated in relation to a player's response to both external/shared rhythms (e.g. percussive music, chanted songs, and the opponent's kinesthetic cadence) and internal/personalized rhythms (e.g. breathing, heartbeat, perceived sounds, and melodic memory). On the one hand, the orchestra's tonal rhythm (*toque*), guided by the leading berimbau, functions as a guiding score or matrix over which the two players at the center build both synchronized and multi-meter interactions. Moreover, since performers are not encouraged to communicate verbally, this rhythmic synchronization provides an additional sensorial channel between the pair at the center and ensures a constant connection between them and the musicians controlling the entire event (circle of capoeira) from the periphery. On the other hand, astute players seek opportunities to come in and out of sync with the main riff in order to disrupt and/or change the pace of the game and potentially gain a dissonant (contrametric) advantage over their opponent.

In addition to their audio-visual perception, capoeira players look for ways to move across the 360-degree surface of the circle while testing their partner's ability to expand their control of the shared space. From capoeira's basic patterns of movement – including side to side, back and forth, or triangular oscillations – players seek to increase their chances of territorializing the capoeira arena, that is, its performative field of action, while maintaining their dynamic sense of coolness. In practical terms, this means that players are constantly circling around each other, re-adjusting their bodies or twisting, bending, turning, leaping, contracting, or extending parts of their bodies, trying to either push the opponent towards the edge of the circle or pull him or her into a trap. In doing so, their call-and-response games also corroborate to limit the other's field of action and/or destabilize his or her temper or mental cool.

According to Mestre Pastinha, "the capoeira player must be calm, tranquil, and analytical" (1964, p. 32). In sum, there are a vast number of variants within which angoleiros employ the ginga aesthetic to push and pull their moving bodies in order to conquer space, gain time, set up traps, play "mind games," ambush the other, combining corporeal dissonance and serious play with coolness, and ultimately stand apart. Some of the fluid, reversible, and polysemic variations with which they control the development of the call-and-response (non-verbal) conversations include, as discussed below: axis oscillations, flow oscillations, size oscillations, level oscillations, distance oscillations, directional oscillations, and rhythmic oscillations.

Capoeira players engage in a variety of *axis oscillations*, including rotations and inversions. On the one hand, rotations refer to the horizontal, diagonal, or tangential turn of the body in its own axis or in relation to another. Rotations may also include the isolated rotation of body parts from adjacent joints, such as the twist of an arm from the shoulder joint. On the other hand, as previously discussed, inversions reference the vertical transfer of weight from the feet to the hands/head or vice versa as the body inscribes a circular pathway on its own axis. It may either be deployed as a transitional point between moves or a sustainable point for several moves. However, it also suggests the "inversion" of a statement, such as transforming a defense into an attack or disarming an act of violence with mockery.

Flow oscillations, which include circular flow and counter-flow, enable players to delineate spiral or curvilinear pathways to dialogue with one another. This facilitates their mirroring symmetry and allows for their uninterrupted motion. A player may move with or against the flow of the opponent while avoiding direct contact. Examples of this circular dynamic are round kicks, spiral moves, and dislocations across the ground (i.e., *rolê*).

Angoleiros may also employ *size oscillations,* such as extension and contraction, in order to alter the overall perception of their bodily size, either contracting or extending their spinal cord and limbs. Inward pathways, for instance, may function as an escape strategy, but may also be useful as a timely infiltration of the opponent's kinesphere. Conversely, outward transitions and wide-open extensions (legs, arms) contribute towards expanding the player's field of action, thus limiting the opponent's range of motion.

In addition, angoleiros articulate *level oscillations* to either shorten or expand the length of their body in relation to the ground as they transfer their weight from one point of support to the other. The alternating use of the hands and feet to support one's body, for example, will push a player to move horizontally closer to the ground. While higher levels present expansive mobility and freedom of expression, lower levels or transitions that utilize three or more points of support on the ground are often safer or more stable situations.

Angoleiros articulate a variety of in-and-out pathways through which they alter the space available between them. These *distance oscillations* are essential to the sustainability of the call-and-response dynamic within a capoeira game and vary according to their intent. Since direct contact is seldom applied, for instance, a player may shorten the distance to his or her opponent momentarily to either intimidate or destabilize the other or to hinder his or her balance. Conversely, movements that widen the distance between players often equate to escapes or changes in direction, giving them wiggle room to breathe, to display virtuoso maneuvers, or, as they put it, to "restart the game." Furthermore, players combine distance oscillations with other variables such as level, size, or axis, in order to execute complex actions such as "to enter-exiting" or "to exit-entering."

Players also articulate *directional oscillations*. As they move, angoleiros are constantly changing their strategic positionality in relation to their opponents in order to escape or avoid the other's controlling gaze, to mislead their adversary, or to briefly slip out of sight. As a player shifts the direction that he or she is facing, for example, the other needs to re-adjust and therefore reconsider his or her strategies. Since this delay creates a gap in time, the person who established the (new) direction assumes control of the game and exercises power over the opponent's actions. Furthermore, players may articulate these relocations in conjunction with (as described above) their ability to "enter-exiting" or "exit-entering."

Finally, capoeira players engage in *rhythmic oscillations*, which enable them to move in and out of sync with their partner. Within a capoeira game, rhythm is neither static nor constant. Similar to their directional oscillations, angoleiros move rhythmically in accordance with their internal cadence, their partner's tempo, and external sounds. Variations or improvisations within the main riff, such as half time, double time, or change of accent (from 1, 2, 3, 4 to 1, 2, 3, 4), may also provoke a displacement in time. This temporal displacement accounts for the development of the game's polyrhythmic tonal patterns. Subsequently, angoleiros tend to anticipate, slow down, or fall out of rhythm temporarily, thus forcing the other player to reconfigure his or her own game flow.

Once a person is able to comprehend how angoleiros string together these and other variant oscillations into call-and-response dialogues, it is worth considering some of the tangible *effects* associated with the ginga aesthetic. One of the most widely employed strategies of persuasion in capoeira angola is *negaça*,[20] a rhetorical artifice that relies on a mixture of fakery and indefiniteness. Grounded in this polycentric and polyrhythmic movement system, angoleiros employ negaça whenever they wish to dissimulate, negate, multiply, or defer unilateral intent and/or signification. Swaying their bodies from one side to the other, for example, players are able to insinuate multiple intentions, but also to obscure or dissimulate them if necessary. Another fairly common strategy is *golpe de vista* (insight maneuver). Employed as an antidote to negaça, this "stare operation" consists of the integrated use of audiovisual perception and muscle memory to foresee, anticipate, or predict the opponent's insinuated and dissimulated intentions and, subsequently, to respond accordingly in a timely fashion. Collectively, these two rhetorical strategies amount to a kind of hide-and-seek game, where the trick or *malícia* consists of mentally visualizing – or premeditating – intentions, emotions, and directions that are being hidden right before one's eyes, while making one's own actions and aspirations seem vague or undecipherable. In other words, within capoeira angola, players counterpoise strategies such as insight with fakery and indefiniteness in order to tilt the asymmetrical negotiations of power relations in their favor, certainly complicating the intent or the *meaning* of these call-and-response dialogues. Subsequently, each strategy of persuasion or rhetoric, including (but not limited to) negaça and golpe de vista, contributes to the orchestration of a balancing act between risk and safety or threat and vulnerability. Moreover, while players swing back and forth between their desire to territorialize the shared space and gain notoriety, they also avoid both direct confrontations and fixed labels.

Here is one example of how the three categories – function, role, and effect – may interact in context. As two capoeira players enter the circle and begin playing a match, they may initially focus on synchronizing their rhythmic flow in order to get to know one another. At this initial stage, they often sway slowly from side to side, avoiding abrupt changes and frictional clashes. Acting like mirroring images, they appear to be dancing harmoniously. As the game develops, they may also employ their dexterity to seductively attract or distract each other and subsequently gain an advantage in the game. This can be achieved, for example, by coming in and out of sync or moving from a synchronized flow to a hide-and-seek chase. At first glance, the dialogue may acquire a ludic or playful quality, where players "suggest" rather than "execute" attacks and "point to" their opponent's open and vulnerable points rather than "take advantage" of them. At this level, humor may become a trump card. Yet, below the surface, players may also gradually craft a subtle but in-depth line of reasoning, like a chess game, punctuated by a series of strategically articulated questions and carefully constructed answers, alongside of which a pair gradually transforms its hide-and-seek maneuvers into a psychosomatic battle over control of the performative space. At this stage, fight-like efforts may be interwoven with both dance-like and play-like qualities of movement in order to build up momentum yet remain calm and focused. Step by step, a player's ability to ephemerally but continuously push his or her opponent into unfavorable positions (the position of predicate or otherness), despite the other's size, shape, strength, gender, social status, etc., guarantees a space at the center that one can call one's own. Similarly, players seek to remain mentally "centered" and emotionally "cool," while "off-centering" their opponent's performative sense of coolness. It is within this ephemeral yet tangible territory that skillful and virtuosic players articulate dynamic negotiations of power-knowledge. By combining technical dexterity and analytical strategies with dazzling acrobatics, for instance, they are able to construct their inter-subjectivity within the public sphere, through improvised actions or, rather, choreographed improvisations. In doing so, they build their own reputations, acquiring respect and admiration, and thus honor, within the circumscribing group. This complex process constitutes one of the defining characteristics of these dancing-fighting-playing-thinking games.

Moving forward

Today it is commonplace to celebrate capoeira as an Afro-Brazilian cultural manifestation. Still, until very recently, very little attention had been given to critically analyze how the capoeira game, i.e. both its repertoire of movements and its underlying logic of organization, reproduce Africanist aesthetic principles. Seeking to overcome such a gap, in this chapter I have offered a set of analytical tools with which to measure and qualify the role of the ginga aesthetic in the practice of capoeira angola. In brief, this movement system acts as an underlying mechanism with which angoleiros re-structure their moving bodies and construct nonverbal dialogues grounded in Afro-centric knowledges, thus acting as an embodied practice of cultural resistance. Through their polycentric and polyrhythmic hip-play or jogo de cintura, angoleiros seek to stand apart, while preserving a particular distinction or judgment of taste connected to Afro-Brazilian heritage. Beyond losers and winners, this practice recuperates-cum-invents a non-hegemonic way of organizing ideas corporeally and moving across time and space, of perceiving and interacting with others, and of constructing one's own processes of self-making in front of a participant audience. Hence, by reconfiguring the way in which one moves across and thinks about the world, capoeira angola cultivates an epistemology beyond colonial languages, whose scope exceeds or differs from Eurocentric thought.

On top of that, circles of capoeira angola may be approached as protective and permissive arenas inside which players haggle, break down, mock, and surprise one another, seeking to tilt or revert the asymmetrical negotiations of power relations in their favor. Moreover, by connecting Africanist principles such as coolness, dissonance, and serious play to concepts such as elegance, beauty, and pride, the call-and-response dialogues that angoleiros improvise inside capoeira circles increase their self-esteem. In the process of constructing flexible choreographies of (self-)identification, players also become active participants in the communities they imagine for themselves.

In particular, this non-hegemonic regime of bodily training and performance enables players to ephemerally exercise a body image of social dignity and civic power, even though several segments of the society continue to view their physical appearance, their socio-economic status, their ethno-cultural values, and their system of beliefs as sources of shame. In doing so, capoeira's choreographed improvisations in general, and the ginga aesthetic in particular, may be understood as non-hegenomic techniques of the self with which to deconstruct, or rather decolonize, the shame, lack of self-esteem, and sense of inferiority inculcated in the players' bodies through colonial mechanisms of power (e.g. the raced division of labor) and effects such as racist discourses. In that sense, both their choreographies of identification and the underlying structure behind those performed actions – the ginga aesthetic – have played an additional role in subverting, or at least questioning, Western/colonial values and regulations of how one should move, behave, stand, or look.

In closing this chapter, I invite the reader to return to Rugendas' historical illustration of a capoeira game. The epicenter of Rugendas' composition, I would like to point out, is occupied by a seated black female figure serving food from a cauldron to a standing black man. Positioned in the back between the two fighters, the acting food vendor and her pleased customer are the only ones whose gaze is not directed to the Africanist war dance. This commercial exchange qualifies the conditions under which these "uncanny" actions take place. Rugendas' carefully constructed composition illustrates how the commercial exchange between female street food vendors (discussed further in the next chapter) and the populace (including slaves for hire) produces territorialized black spaces at colonial urban sites in Brazil. Though subtle, the female food vendor's informal trading spot or "corner" (*canto*) authorizes, i.e. protects and permits, the execution of an Africanist enactment of sociability by individuals that are otherwise living in a "state of exception" (Agamben, 2005).

5
What is it about the Baiana?

Life is a Marketplace and our true house is in heaven.
Yoruba proverb

Introduction

In 1939, the notable Carmen Miranda starred in the Brazilian motion picture *Banana da Terra* (literally *Homeland Banana*), performing the – now historical – Dorival Caymmi's samba "O Que é Que a Baiana Tem?" ("What is it about the Baiana?").[1] Briefly, the song pays homage to the *baiana*[2] (i.e. a black female ambulatory vendor) who "typically" sells Afro-Brazilian food on the streets of urban centers. Miranda's impersonation of this market woman within a modern casino show illustrates, more importantly, the enduring popularity of this character-type within Rio de Janeiro's entertainment industry. Accompanied by a group of samba musicians, *Bando da Lua*, Miranda sings:

What is it about the baiana? 2x
[Literally, "What is it that the baiana has?"]
She has a silk turban. (Yes, she) has.
She has gold earrings. (Yes, she) has.
She has gold bracelets. (Yes, she) has.
She has *pano da costa*.[3] (Yes, she) has.
She has an embroidered blouse. (Yes, she) has.
She has gold necklaces. (Yes, she) has.
She has a starched skirt. (Yes, she) has.
She has dressed-up sandals. (Yes, she) has. (Author's translation)

At first glance, it might seem that the lavish garments and ornaments with which the baiana adorns her body define her raison d'etre. Yet, as Miranda's performance progresses on the silver screen, it becomes clear that the "thing" that she *has* (i.e. the most precious attribute of this character) is not her "typical" baiana attire. So, what is it about the baiana? As Caymmi's lyrics indicate and Miranda's dance insinuates, the baiana's inherent drawing power comes from the syncopated way her body (especially her hips) moves to the black rhythm of samba – in simple words, "how well she shakes it" (*como ela requebra bem*). Despite her white-looking skin, Miranda's subtle yet sensual hip-swings authenticate the Africanicity of the baiana figure she impersonates.

At that time, Miranda was already a household name within Rio de Janeiro's entertainment industry. Yet, once she adopted this dancing act as her signature style, the Portuguese radio singer/actress who grew up in Brazil became an overnight "bombshell" sensation. Miranda's burlesque impersonation of the baiana figure, an ambiguous enactment of femininity that looks white but acts black, finds a compelling resonance with the *democracy of races* ideology that gained momentum during Vargas' Estado Novo (1937–45). In short, Miranda's live and mass-mediated performances contributes to *fixing* this carnivalesque samba dancing baiana as one of the most forceful, yet controversial, performances of national identification ever produced about Brazil. Like a shooting star, in 1940, Miranda landed in Hollywood, where her extravagant character was hauled into betokening, on global stages, the personification of exotic otherness, the "South American Way." Finally, Miranda's extraordinary charisma and her mass-mediated visibility across the 1940s and 1950s crystallizes the baiana figure as a fixed idea.

In strong contrast with this set choreography of identification, in this chapter I deconstruct "the baiana" as a stock-still idea bound up with a once-marginalized dancing rhythm (samba) and a national holiday (carnival). In order to accomplish this, below I retrace a genealogy of the baiana figure, from "real" black female food vendors across the Portuguese colonial empire to "fake" ones, such as Miranda's interpretation of "The Lady in the Tutti Frutti Hat" in Busby Berkeley's *The Gang's All Here* (1943). In my genealogy, I first identify a general network of roles that black female food vendors have performed since colonial times and the power relations they articulated, in particular, on the streets of Brazilian port cities such as Salvador and Rio de Janeiro. Next, I investigate two distinct spheres, where this figure gained wider visibility, and renewed dimensions, following Brazil's late abolition of slavery (1888).

I look at Rio de Janeiro's revue theatre (*teatro de revista*) and its black carnival pageant-parades (*ranchos de carnaval*). As I will demonstrate, men and women from different classes and ethnic backgrounds have long been appropriating the image of (colonial) market women as a floating signifier with which to articulate dissimilar ideas, from mockery to self-assertion.

In concluding, I return to Carmen Miranda's immortalized impersonation of the *fake baiana* in order to investigate how this complex figuration becomes an "authentic" symbol of Brazilianness during the Estado Novo. The character that Miranda personifies at the end of the 1930s, I propose, should not be taken as a direct representation of historical Afro-Brazilian female street vendors. Instead, it constitutes a bricolage of citations and samplings of discourses circulating in Brazil since colonial times, including references to showgirls, sex workers, drag queens, and black mothers (or Mama Africa). Meanwhile, the election of a figure that looks white but acts black to represent Brazil's "local flavor" reaffirms the country's mixed feelings about its slavery past and its miscegenated present. In that sense, Miranda's hyperbolic character and the mass-mediation of her set choreography perpetuates and enlivens, rather than resolves, such a conundrum (see the Introduction). Like the tip of an iceberg, the modern success of Miranda's fake and indefinite rendition of the baiana figure sits on top of a much deeper paradox particular of Brazil's coloniality of power: the (already mentioned) sexual interaction between races inside a community that was unable to dismantle its racial division of labor in modern times. In the end, Miranda's modern commoditization of this set of long-familiar practices and the shallow extravagance of the persona she embodies repeatedly had the effect of sterilizing and freezing the baiana figure, thus limiting its power of negotiation.

Below, I attempt to look at what is behind the character that made Carmen Miranda famous internationally. Guided by the question "What is it about the baiana?", my excavation below is an effort to remember the tightrope along which the baiana figure has walked between extreme positions of pride and shame and the balancing acts it has choreographed alongside its path. My analysis also seeks to question how, underneath its looks, garments or props, its way of moving and its socio-economic visibility might have either expanded or constrained the mobility of marked bodies in Brazil. The power of this choreographed process of identification that I call the baiana figure lies in its ability to construct desire, thus attracting the viewer, while deferring static significations. The baiana figure swings across different

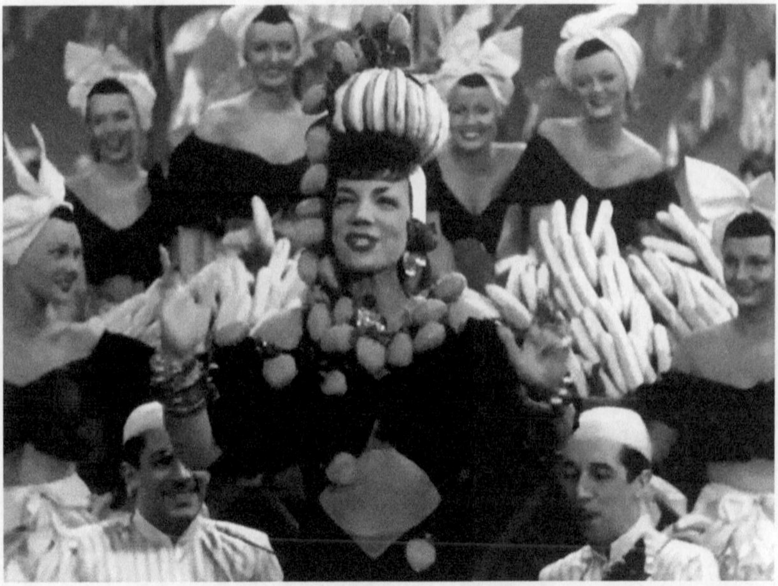

Figure 5.1 "The Lady in the Tutti Frutti Hat." Screenshot of Carmen Miranda in Busby Berkeley's musical *The Gang's All Here* (1943)

scenarios. In doing so, the semantics attached to its improvised performances oscillates between pre-conceived ideas of otherness and polysemic interpretations of selfhood. Beyond the mainstream codification of the *baiana type* as a fixed idea, the ephemeral acts it performs across workdays and holidays have the potential to improvise faults and seams that ephemerally discontinue the linear flow of marked roles. Its metaphoric jogo de cintura – a juggling of ideas that bargains accessibility while deflecting prejudices – seduces more than signifies: it alludes and eludes meaning in different directions. Hence, my narrative below should not be read as the reduction of its realization, but rather as my attempt to improvise a dialogue with what it *does* on top of what it *looks like* or whom it might *represent*. The baiana figure articulates, in my text, *flexible choreographies of identification*.

Instead of ironing out the differences between "fake" and "real" baianas, this chapter seeks to elucidate, on the one hand, how this once widely available yet twice othered figure evocative of African ancestry in Brazil employs fakery and indefiniteness to negotiate inter-subjectivity from a place of otherness. On the other hand, I also investigate the means through which the baiana figure eventually gains renewed

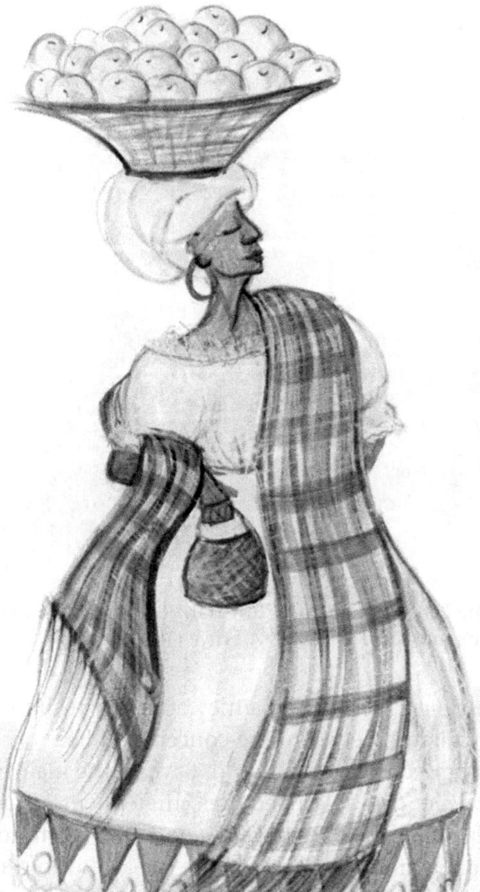

Figure 5.2 Colonial market woman (*quitandeira*) © Cristina Rosa, 2014

currency across orchestrated enactments of revelry (carnival parades and burlesque musicals) and, later on, is employed to concoct discourses of national identification. Still, as the anthropologist and music scholar Hermano Vianna problematizes in *The Mystery of Samba*: "How does one define a culture invented (imagined or projected) on the principle of indefinition?" (1999, p. 109).

Historicizing the baiana figure

Ambulatory black female vendors specialized in the trade of aliments constitute one of the most widely depicted figures within colonial illustrations of urban centers across the Lusitanian (Portuguese-speaking) world. Food vending was not forbidden to men. Yet, as historians seem to agree, women dominated the commercialization of products such as raw, dry, and homemade cooked food, long before the arrival of European merchants and colonizers in Central-West Africa. In the region of today's Angola, for instance, female vendors working in public areas such as open fairs (*kitandas* in Kikongo) exercised an active role within their local economy. Nevertheless, with the development of plantation systems in the Americas, the scope of their trading relations tended to shift. The accelerated trafficking in "human flesh" across the Atlantic Ocean, as Joseph Roach (1996) puts it, contributed to rearranging, in particular, the role that these local merchants occupied and the functions they fulfilled within their own communities. After the institution of the transatlantic slave trade system, the African-Portuguese term *quitanda* was extended to indicate marketplaces within Portuguese-controlled port cities where anything from salt to human beings was sold or exchanged. In these communities, market women, commonly called *quitandeiras*, assumed the job of stocking European black ships with non-perishable items such as flour and dried fish, in addition to feeding those involved in that transatlantic enterprise. Given the coloniality of power established in these African port cities, many local market women were mistaken for sex workers or were coerced into also providing such services. As the Angolan poet António Agostinho Neto[4] suggests in his poem *A Quitandeira* (1974), "A quitandeira / que vende fruta / vende-se" ("The market woman / who sells fruit / sells herself").[5] As Agostinho Neto suggests, this dehumanizing trading system and its unequal power relations have forever loaded the term "quitandeira." It now carried the colonial marks of violence and exploitation over the bodies of African women and – by extension – the bodies of their offspring trafficked into the New World.

In colonial Brazil, on the other "margin" of the Atlantic Ocean, the term "quitandeira" came to signify, quite specifically, black women, initially all slaves for hire (*escravos de ganho*),[6] who sold fruits, vegetables, and cooked food on the streets of urban centers. These market women became particularly famous for the African-inspired sweetmeats, stews, and porridges that they sold to the population at large. Food vending constituted, more importantly, one of the primary ways that these slaves for hire were able to purchase their own freedom. These "breadwinners" (*ganhadeiras*) worked long hours and most remained poor. Yet, there were those who not only accumulated enough money to emancipate themselves (and their relatives), but were eventually able to establish a name for themselves, running autonomous and relatively profitable businesses. As the work of traveling artists and explorers illustrates, it was not uncommon to glance at a quitandeira adorned in imported fine lace and golden jewelry during the slavery era in Brazil. Nonetheless, as their businesses expanded, many also employed freed and enslaved labor (see Figure 5.3).

These street vendors were known for their precious garments and adornments and their large baskets or trays (*tabuleiros*),[7] on which they could carry their entire business set-up. They were also known for their languid and offbeat swayed-walk through which they balanced such trays on their heads. Generally speaking, these "tray women," as they were often called, employed the ginga aesthetic as they walked, for gliding their hips from side to side helped them to elegantly carry things on their heads and free their arms. While their hips remained the control

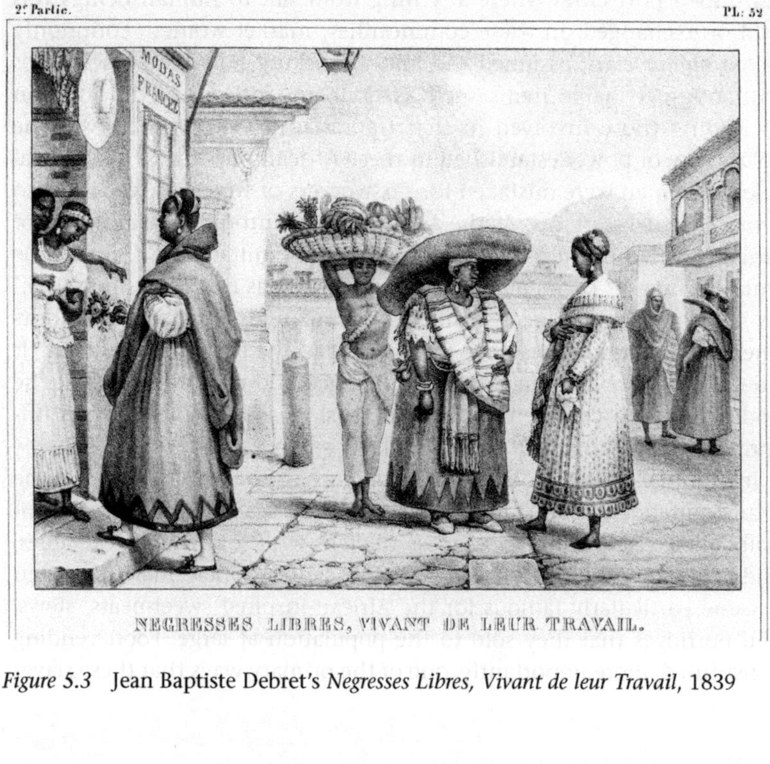

Figure 5.3 Jean Baptiste Debret's *Negresses Libres, Vivant de leur Travail*, 1839

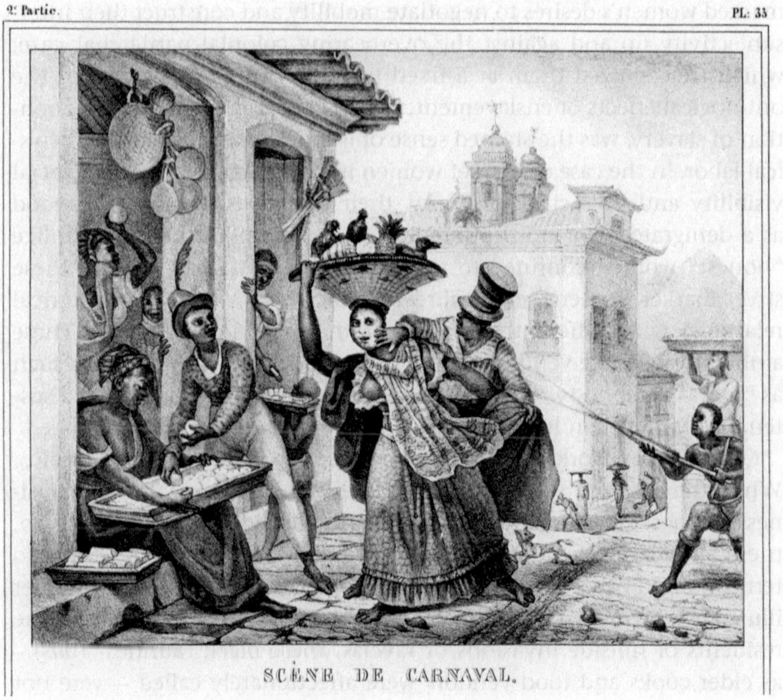

Figure 5.4 Jean Baptiste Debret's *Scène de Carnaval*, 1839

center of their balancing acts, their flexible spines worked like luscious shock absorbers counteracting the weight of their large baskets or wide trays with subtle undulations. On top of this practical function, their swayed-walk also infused the sweat of their laboring effort with ethno-cultural meanings. Reversibly, the grace and auspiciousness with which these market women moved around town drew the attention of their customers towards their selling goods and authenticated the processes of self-fashioning they negotiated on the streets. The success of their commercial enterprise also projected a kind of dignity and a sense of pride towards their choreographed blackness, (i.e. their *gingado*).

European traveling artists have drawn one's attention to the (unconventional) wealth and freedom with which some of these black female workers carried themselves in society. Yet, above the trays they balanced on their heads, these street workers carried a complex and unresolved load specific to Brazil's process of colonization. A deep-rooted tension tangling both gendered and racial divisions of labor pushed these marked women's desires to negotiate mobility and construct their intersubjectivity up and against the overbearing colonial patriarchal gaze, which (re)cognized them at a fixed position of otherness. One of the ontological effects of enslavement, which persisted even after the abolition of slavery, was the stained sense of inferiority associated with physical labor. In the case of market women in particular, despite their social visibility and financial autonomy, their work was locally understood as a denigrated – or not honorable – occupation. Furthermore, unlike "honest" women confined to the privacy of patriarchal homes, these street market women were in direct contact (although in asymmetrical relations) with different segments of that patriarchal society. These ambulatory female vendors tended to be associated with concepts such as "moral dishonesty" and "social degeneration," even if they sold sustenance rather than lust.

Quitandeiras worked either independently or in organized groups. While they walked with trays on their heads, many set up their business on a set location at markets, parks, or sidewalks. In doing so, these ambulatory vendors transformed public sections of the city into territorialized black spaces, as the capoeira print in the previous chapter illustrates (Figure 4.1). Ephemerality is key to their process. Unlike the residents of hillside invasions or favelas, these black "aunties" (*tias*) – as elder cooks and food vendors were affectionately called – were not vested in taking permanent ownership of state-owned land. Yet, it is possible to establish a relationship between the public territories these working matrons occupied daily, their financial independence, and the sense of ethnic pride they extended to their kinswomen.

Following the decline of the plantation economy in the northeast and Brazil's gradual process of the abolition of slavery (1850–88), urban centers in the southeast became primary destinations for freed slaves leaving rural areas. Afro-Brazilian individuals and families migrated especially to Rio de Janeiro – Brazil's capital since 1763 – seeking to find better job opportunities and build new kinds of communities for themselves. This intra-migratory movement led to an historical process of re-Africanization of urban black sociability and religiosity in that city. As addressed in Chapter 2, during the First Republic (1889–1930), Rio's European-aspiring elites developed a resentful awareness of the ways in which local black customs and behaviors had permeated and continued to penetrate their post-colonial lives. In this "marvelous city" (*cidade maravilhosa*), black subjectivity became an idea out of place.

Similar to the "fear of capoeira," mainstream society tended to view the widespread proliferation of black food vendors in post-colonial

Brazil as a social disease, especially in places where they had continuously *territorialized* sections of the city. At the time, Schwarcz (1993) proposes that local authorities, along with medical and legal magistrates, adopted ideologies such as scientific racism to combat or at least contain the city's "infectious" display of blackness. And, since hygiene constituted one of the practical concerns regarding the informal commerce of food in public areas, these black food vendors were often blamed for or mixed with the bad odor and the promiscuity disgracing Brazil's capital. This ethno-cultural battle reached a climax in the 1910s, when Mayor Pereira Passos installed an aggressive campaign to civilize/de-Africanize the city. His gentrifying measures included the demolition of the downtown slum tenement-houses (*cortiços*) to pave the way for European-inspired boulevards and modern edifices, the relocation of those poor (mostly non-white) residents to the outskirts, in neighborhoods such as Saúde and Cidade Nova, and the generalized persecution of black street figures such as quitandeiras, capoeira players, and batuque revelers.

Despite avid persecution, these market women's informal but long-lived presence and trading relations remained as one of the few street occupations whose display of Africanicity resisted (or strategically survived) the structural transformations set in motion during Passos' "get out" policy. Unlike capoeira players and batuque revelers, mostly non-white men historically associated with violence, roguery, or idleness, *quitandeiras* were working-class women whose choreographed actions projected an image of sensuality, independence, and self-sustainability. Unlike their male counterparts, these black and mulatto women used their profitable occupation to bargain a space they could call their own. Through their working journeys, these businesswomen constructed a choreographed presence in the New World. They also wore colored pano da costa laid over one shoulder, which signaled their ethnic background and/or their social status in relation to Africa, and metal rings, bracelets, and necklaces, through which they displayed the wealth they had acquired in the diaspora. Both the territories they occupied and their strategies of negotiation in the public sphere brought them self-esteem. Their socio-economic mobility thus enhanced the processes of self-identification they articulated from a marginalized position.

Underneath their African-inspired costumes and goods, these market women's way of moving and negotiating visibility were, above all, the means through which they claimed their ethno-cultural logic (their mode of self-fashioning and interacting with others). The ginga aesthetic structured, in particular, their articulation of femaleness and blackness. Moreover, these socio-economically active matrons and their everyday life performances on the streets fostered the emergence of a positive and coherent image of and for Afro-Brazilians, even though gendered and racial roles were not subverted.

The nineteenth-century migration of Africans and Afro-Brazilians from the northeast to the southeast brought about a fierce competition between the local and migrating *quitandeiras*. Their rivalry was quite visible as this profession had acquired regional trends over time. In Rio de Janeiro, local market women incorporated black fabric in their wardrobes, most likely influenced by Portuguese fashion and etiquette. In contrast, many vendors coming from Bahia, Brazil's colonial capital, distinguished themselves by their extensive use of white garments, which was indicative of their affiliation to Candomblé and other Afro-Brazilian religions that emerge in the second part of the nineteenth century. In Bahia, the flexibility of street vending became an ideal line of work for those committed to their temple's reclusive and time-consuming activities. For instance, many priestesses and devotees of the *orixá* Oya/Iansã (see Figure 2.3), a female warrior deity from Candomblé Keto associated with marketplaces and the Niger River (*Odo-Oya* in Yoruba), specialized in the commerce of *acarajé*,[8] a votive dish associated with that entity. In addition, these nicknamed "Bahian women," or *baianas*, made extensive use of sacred prayer amulets (*patuás*), religious beaded necklaces (*guias*), and *balangandãs*, a chain of protective mementos cast in metal. Attached to their waist by a large pin, the baianas' balangandãs made a rattling sound as they moved with their swayed-walk. Along with their lavishing attires and adornments, the aesthetic knowledges these ambulatory food vendors articulated in their everyday life performances – including the coolness of their swayed-walk, made them fascinating and memorable.

Gradually, the terms "quitandeira" and "baiana" became synonyms within the streets of Rio de Janeiro and, by the 1920s, the latter became the predominant one. Though often ridiculed, persecuted, or exoticized, in the end the *baiana figure* became a popular leitmotif within the collective imagination of Brazil's modern capital. Yet, this conversion in name – from the African-Portuguese quitandeira to the Brazilian-Portuguese baiana – does not necessarily reflect the large migration of female black vendors from one particular geographical location (Bahia) and their dominance in Brazil's capital. Rather, it reveals a pejorative association that the adjective "Bahian" (baiana or baiano) acquired in Rio de Janeiro, the "possible Europe," as a euphemism for African heritage or, for that matter, anyone looking or acting black. At the same time, ambulatory food vendors from different regions also adopted the typical "Bahian" attire as a marketing strategy to further authenticate the Africanicity of their selling goods in Rio de Janeiro. In this scenario, the "Bahian woman" who sold food dressed in a colonial attire adorned with Afro-Brazilian religious symbols (re-)emerged as a character-type, which illustrated or recalled Africa, colonial past, black motherhood, and auto-exotic otherness.

In spite of how they were perceived by the ruling classes, black food vendors in Rio – or "real" baianas – continued to cultivate a sense of pride around their Afro-Brazilian fashion, ornaments, recipes, manners, gestures, and beliefs. And they certainly had to evade or at least attenuate the epistemic violence and the endemic shame under which their neglected and overworked bodies had been recognized.

Many of the "Bahian aunties" (*tias baianas*) who staged those balancing acts on Rio's streets also played other relevant and interconnecting roles within the socio-political life of their marginalized communities. Historical "aunties" such as Tia Ciata (see below), Tia Perciliana, Tia Amelia, and Tia Bebina are amongst the legendary black mother-like figures who migrated to Rio and populated the neighborhood of Cidade Nova, nicknamed "Little Africa" (*Pequena África*)[9] after Passos' gentrification of the downtown. The social roles they accrued beyond their working journeys also shaped these market women's syncopated way of moving. In fact, it is through their active participation in these accumulating activities discussed below that they learned to articulate their individualized swayed-walk or swagger. Their interconnecting social roles offered, in that sense, valuable departure points from which Afro-Brazilian communities could re-imagine and valorize their collective singularity.

First and foremost, it is worthwhile re-stating that many tias baianas were actively involved in Afro-Brazilian religions. As *baianas do Candomblé* or priestesses, they were responsible for ensuring the success of their private celebrations and the transmission of embodied knowledges. In that capacity, the sacred dances and ceremonies they performed contributed to recuperate-cum-invent their Africanist way of moving (i.e. the ginga aesthetic). As I will address below, many food vendors such as Tia Ciata were also known as *baianas festeiras* or party organizers. Their famous house parties at Little Africa offered a family atmosphere and a lived repertoire of selective inventiveness, especially in the kitchen and on the dance floor. In doing so, they provided both material and affective sustenance with which to feed and renew the ethno-cultural processes of identification of their entire community. Besides live entertainment by local artists, these "underground" dancing parties also featured a wide range of guests coming from different places, classes, and ethnic backgrounds. Thus, their house parties functioned as *cultural trading posts* across distinct kinds of peoples and ideas. In the long run, the house of these festive matrons became a fertile ground for Rio de Janeiro's local black music scene of the time (see Chapter 3) and, subsequently, many sons and grandsons of these tias ended up as notable samba composers with record deals.[10]

Many of these distinguished market women lived around *Praça Onze*, a public square within Cidade Nova where the "Bahian colony" had settled.[11] Due to their location, some of these matrons opened their houses to newly migrating families, offering them shelter until they could find a place to live and work in the capital. In that capacity, these tias also acted as *surrogate mothers*, providing displaced and marginalized individuals with a space to settle and make new connections, forming coherent groups from displaced or migratory individuals. Furthermore, as *baianas da sociedade* or socialites, successful market women cultivated a series of external connections and "favors" in order to, for instance, acquire police permits to execute their marginalized celebrations within "Little Africa" or find someone a new job.

In the end, the territories these mother-like black figures occupied and the positions they negotiated daily with society at large as food vendors i.e. their eating habits, their cooking skills, their social etiquette, their religious beliefs, and their distinction of taste, permeated across the accumulating ways through which they interacted with friends, neighbors, party guests, and immigrants. Collectively, Rio de Janeiro's *tias baianas* came forth, through the multiple positions they exercised, as both a bio-ethnic signifier and a site from which people in "Little Africa" could juxtapose and reformulate ideas of and about themselves and their immediate community. In this sense, concludes Monica Velloso, even their bodies translated the idea of a territory (1990, p. 18).

Nevertheless, it is important to clarify that the multiple roles branching out of this black figure in the New World carry no ontological essence. The multiple significations that these tias baianas have evoked ephemerally through their actions, rhythms, garments, and merchandise vivified pliable and transient processes of identity formation in post-colonial Brazil. Their overlapping choreographies of gender, race, class, place of birth, religious affiliations, etc. were contingent upon their shared knowledge and individualized inventiveness as well as the socio-historical terms of their local(ized) productions. Hence, above the pseudo-authenticity that the so-called "real" baianas disseminated, these archetypical black "aunties" choreographed efficacious discourses of self-identification across private and public spheres or between leisure and work environments. Through playful re-fashioning and tasteful imagination, their multiplying choreographies of identification (e.g. baiana de Candomblé, baiana festeira) have collectively functioned as rituals for re-organizing the aesthetic knowledges and the cultural memory of Afro-Brazilians within their newly imagined communities.[12]

The larger-than-life Tia Ciata, Hilaria Batista da Almeida (1854–1929), stands as the primary (though not the only) example of these historical – but locally mystified – tias baianas. Born in the countryside of Bahia, Tia Ciata was initiated in Candomblé and exposed to samba de roda gatherings of that region, as well as the religious celebrations by the Afro-Catholic Sisterhood of Boa Morte (see Chapter 3). According to Moura (1983), Ciata moved to Rio de Janeiro at the age of 22 and eventually married João Batista da Silva, an affluent Afro-Brazilian. In Rio, Ciata established a business specializing in selling Afro-Brazilian sweetmeats that expanded over time. Her franchise hired local neighborhood women to sell her merchandise across town, dressed in the typical baiana attire. Beyond her profitable business, at the turn of the twentieth century she assumed the leadership of the (now) historical *Rosa Branca* (White Rose), one of the first black Mardi Gras krewes (ranchos) to parade on the streets of Rio de Janeiro in the wake of its prohibitions (see below).

Tia Ciata's culminating contribution to Rio de Janeiro's "black music" scene was, nevertheless, her famous house dancing parties. Her dancing events offered multiple environments with distinct musical genres (e.g. batuque, samba, maxixe, choro) and the guest list included a multitude of local and foreign musicians, dancers, artists, writers, intellectuals, journalists, and even politicians (Moura, 1983; Chasteen, 2004). The police permits for her gatherings were often granted with the help of her husband's professional connections and, over time, these social events served as rehearsals for subsequent public enactments of cultural revelry and resistance for/of local blacks and mulattoes in Brazil's capital. Given Tia Ciata's importance to the Cidade Nova community and the interconnected historiographies of samba and carnival in Rio de Janeiro, her life's work has been evoked within the literature of her time – e.g. Mario de Andrade's *Macunaima* (1928) – and brought back to life through endless songs and anecdotes.

To be clear, despite the socio-cultural contributions that "real baianas" such as Tia Ciata articulated in Rio de Janeiro's "Little Africa" and the working interactions they established with that post-colonial society at large, in their eyes, black female bodies continued to be twice othered. In the end, though, the multi-faceted and fascinating figure of the baiana seduced and, to a certain extent, "contaminated" the collective imagination of Brazil's capital.

Modern representations of the baiana figure

In the first part of this chapter I recalled the multiple roles that "real" market women performed in port cities across the former Lusitanian Empire. Nevertheless, images of these ambulatory food vendors also resurface on both sides of the Atlantic as a character type represented and/or impersonated in different contexts. For instance, according to Selma Pantoja (2000, 2001), "the kitandeira" was a popular costume during pre-Lent festivities in Angola since colonial times. In places such as Luanda, men took advantage of these "festivals of inversions" to playfully cross-dress as market women, often ridiculing or exaggerating their attributes. Likewise, Rio de Janeiro's quitandeiras and baianas had been adopted as a popular motif during carnival, especially amongst men, within elitist processions. Outside Mardi Gras, "the baiana" reappears as a stock character in a variety of proto-Catholic pageant-parades or *folias* (see Chapter 2).

At the end of the nineteenth century, impersonations of these black market women began to pop up in different cultural productions across Rio de Janeiro. Below, I analyze the emergence of the baiana figure in two distinct contexts. On the one hand, as I will demonstrate, the baiana figure emerged as a staple character type on local burlesque stages known as *teatro de revista*. At these musical acts, European actresses danced to Afro-Brazilian music dressed as "lascivious" market women. On the other hand, the baiana figure assumed a new centralized position within dignifying street parades put together by black neighborhood organizations. Later on, these street processions evolved into *samba school parades*. Certainly, staged musicals and street parades are not the only two realms where the figure of the baiana resurfaces at the turn of the twentieth century. Nevertheless, these two heterogeneous contexts allow me to examine different processes of translations (and recodifications) of the "reality" of local baianas in post-colonial Brazil. By comparing and contrasting them, I also note how the baiana figure connects to other processes of ethno-cultural resistance, as the country transitioned from an agricultural monarchy (1808–89) anchored by (African) enslaved labor to a post-colonial republic (1889–1930) supported by a (European) wage labor force.

Samba school parades

Before I address how Afro-Brazilian communities re-inserted the baiana figure into dignifying enactments of revelry during Rio de Janeiro's street carnival at the end of the nineteenth century, here is what you need to know. Since the transference of the Portuguese court to Brazil in 1808, which transformed Rio de Janeiro in the capital of that empire, local authorities had repressed, and nearly extinguished, street carnival festivities (*entrudos*) in the capital due to their excessive violence and "savagery." The populace was limited to celebrate this and other holidays in the outskirts of the city. Around the 1850s, exceptions were given to the Great Carnival Societies, exclusively male elitist organizations, whose processions emulated the French and Italian carnival processions. Accompanied by the sound of operas, these high-society organizations paraded through the affluent Ouvidor Street on horseback, large floats, and, later on, automobiles. Their European-aspiring parades combined, amongst others, beautiful actresses representing political themes and ideas such as democracy and liberty, with men grotesquely cross-dressed as non-white female street workers, somewhere between food vendors and prostitutes.

Towards the end of the nineteenth century, this scenario began to change. First, the urban petit bourgeoisie started to also parade in semi-organized ranchos. Unlike the Great Carnaval Societies, they paraded with smaller floats and less elaborate costumes to the sound of marching bands. In around the 1880s, black ranchos such as the *Cucumbys*, a pageant-parade similar to colonial Congadas, emerged. As N. Lopes (2005) explains, they figured out a way to participate in Rio's carnival by first applying for a permit to parade during the Epiphany cycle festivities. Then, whenever the syncopated music of their black-only folias and their lascivious ways of dancing came into conflict with the Catholic Church's standards, they asked to transfer the date of their street performances from Epiphany to Mardi Gras. This strategy was shortly followed by other groups.

At the turn of the twentieth century, legendary carnival ranchos such as Tia Ciata's Rosa Branca (White Rose) and Jovino Hilario's Rei de Ouro (King of Diamonds) banked on the glorifying pomp, frame, and characters of these proto-Catholic pageant-parades to restore a sense of (black) pride and dignity towards their urban communities and their marginalized culture. In doing so, these Afro-Brazilian working-class processions adopted the structural elegance and "honesty" of Epiphany ranchos to counteract the disrepute attributed to their local black aesthetics. Like their colonial counterparts, these ex-Epiphany ranchos were sponsored by community-based organizations that functioned year-round and galvanized a considerable amount of labor and devotion, if not money, towards their annual productions. Like the burlesque revistas (see below), these lavish processions did not follow a coherent theme or tradition. Instead, they were structured as a kaleidoscope of juxtaposed elements from different parts of the world, which might include Japanese lanterns, Amerindian feathered headdresses, West African instruments, Venetian floats, and local dancing rhythms.

These "modern" black carnival parades were composed largely of a pack of beautified women, nicknamed "pastoras," followed or led by a marching band. This format followed, one should note, the molds of Profane Nativity Scene parades (*Pastoril Profano*), a particular kind of folia popular across the northeast plantation regions of Brazil since colonial times. While the "religious" Nativity Scene parades focused mostly on the biblical birth of the baby Jesus, it should be noted that its "profane" version featured a pack of naughty dancing "shepherdesses" (*pastoras*) who wiggled their hips on stages or across towns in exchange for church donations. Moreover, modern black ranchos such as Tia Ciata's notorious Rosa Branca featured carnival pastoras dressed in costumes similar to those her daytime employees, or "real" baianas, wore year-around. Due to the popularity of this thematic exchange, female revelers dressed as food vendors or carnival pastoras would eventually be nicknamed "fake baianas" (*falsa baianas*). Nevertheless, contrary to the elites' ridiculed or fetishized representations of black food vendors, the fake baianas at these Afro-Brazilian parades forged a positive connection across the dignity of these market women and the sensuality of their Afro-Brazilian way of moving.

At these carnivalesque pageant-parades, the baiana figure resurfaces as a renewed choreography of collective identification that asserts black pride and self-esteem from a place of otherness. In other words, while the "reality" of market women continued to be fixed according to gendered and racial divisions of labor, the hyperbolic discourses that fake baianas choreographed once a year inside these community-organized processions (as well as other unstructured carnival celebrations) empowered the group with a renewed range of possibilities for representing themselves otherwise. These auspicious packs of young black female dancers paraded across Rio's Ouvidor Street and danced in front of local newspaper headquarters with their lavish costumes, their rattling jewelry, and their "scandalous" thrusts of hips-and-thighs (*requebrado*). To give an idea of the repercussion that these transgressive acts produced within that mainstream society's judgment of taste, in 1900, the newspaper *Jornal do Brazil* praised Rosa Branca's pastoras as a "multitude of charming and beautiful baianas dancing samba" (quoted in Chasteen, 2004, p. 45).

At the *Rei de Ouro* and *Rosa Branca* ranchos, in particular, fake baianas performed ephemeral acts of revelry-revolt against the epistemic racism with which their dark-skinned bodies had been codified in Brazil. Over time, the baiana-pastora leitmotif, or sexy dancing market woman, became an inventive strategy of cultural resistance. Under this "flag," the Afro-Brazilian communities in Rio de Janeiro re-appropriated the public (festive) space to articulate multiple and inventive discourses of self-fashioning. In the case of hip-wiggling baianas portrayed at black parades, their dancing acts infused their traumatic memory with joyful and humorous experiences. The sensorial pleasure of parading across upscale streets in lavishing costumes, during these festivals of inversion (carnival), constituted the common denominator binding that ephemeral mechanism of cultural resistance.

Teatro de revista

In the previous section, I highlighted how Rio's local black community appropriated the honoring structure of Epiphany ranchos to orchestrate renewed street parades during carnival. There, the baiana figure and its sensual way of moving was honored and celebrated. Most significantly, these non-hegemonic choreographies of self-identification tied blackness to beauty, pleasure, and dignity. Below, I examine the inclusion of the baiana figure in a radically different scenario: the *teatro de revista* (revue theatre). Popular in Rio de Janeiro from the 1880s to the 1950s, teatro de revista is a light-hearted theatrical genre that follows the Portuguese "revue of the year." At first, Brazilian revues re-staged the primary events of that nation in the making, from politics to the arts, mixing facts with farce. Early revistas moved randomly from epidemic outbreaks and economic crisis to juicy scandals and circus performances, in short sketches filled with stock characters, allegories, personifications, caricatures, and hasty metaphors. Over time, Rio's revistas increased their musical scenes, capitalizing on young showgirls, often European immigrants, performing tantalizing dancing acts. Nicknamed the "hip-shaking theatre" (*teatro do rebolado*), revistas constructed a fascinating and overwhelming world of fantasy, playfulness, and (available) sexuality catered largely to a male audience (Lopes, 2001). At the high of its effervescence, revistas (re-)created Brazil as an exotic locale infused with innate roguery and shameless women. At the revistas, everything was resolved with, or culminated in, a carnivalesque lasciviousness.

At the end of the nineteenth century, the producers of revistas stumbled upon a "magical recipe" that guaranteed the success of many musicals to come: a white-looking actress dressed as a black market woman, singing and dancing to local rhythms infused with Afro-Brazilian musical cells. The earliest example of this miraculous "mixture" appeared, according to Gomes and Seigel's research (Gomes and Seigel, 2002 and Seigel and Gomes, 2002), in a musical that premiered right after the fall of the monarchy and the institution of a republican regime (1889). Dressed as a charming quitandeira, the young Greek soprano Ana Menarezzi interprets the Brazilian tango "Sabina das Laranjas" (Sabine of Oranges) in the revista "A República" (The Republic, 1890). Regarded as one of the most memorable theatrical scenes of that century, this musical number was in fact a spoof of a real incident involving a food vendor, a parade, and a group of medical students. As Gomes and Seigel explain, in the previous year, the Imperial police had removed an elderly orange vendor from her trading spot near Rio de Janeiro's School of Medicine, impounding her merchandise. Following the incident, a group of republican medical students organized a parade. The reported procession was not a humanistic uprising in defense of a poor old orange vendor. Given the circumstances, it was most likely a mock-protest against the Portuguese monarchy. Yet, when that story is brought into the limelight, the revista transforms the elderly worker into a young and likable *mulata*.[13] The quitandeira's music, her costume, and her dance all referenced blackness. Meanwhile, Menarezzi's soprano voice and her European features and manners performed a desirable whitewashing of the character she staged. Newspaper reviews of the time also confirm that the lasciviousness of Menarezzi's crossover performance drove the local (white male) audience into a frenzy. Following its success on Rio's stages, the contaminating appeal of this hit-song overflowed into the following year's street carnival, landing on the mouths of singing revelers, thus influencing their (judgment of) taste. In 1902, this ebullient composition was included in the first album recorded in Brazil, this time under the title "Laranjas da Sabina" (Sabine's Oranges).[14]

Given its broad appeal, the mixed-race dancing quitandeira was rapidly incorporated in other musical plays. As a stock character, this racially undefined and auto-exoticized market woman fulfilled the same fixed gendered roles common to French cabaret musicals. Yet, contrary to other character types common in European musicals, revista's sexy baiana-mulata formula and the multiplying baiana-pastoras emerging in Rio's carnival celebrations fed off each other. Indirectly, they both resembled the colonial "naughty shepherdess" character type from the above-mentioned Pastoril Profano. Unlike any of them, the revistas' tangible mixture of colonial-patriarchal desires with the aesthetics of modern musicals transformed these white-looking "sexy mulatas" into an oscillating fetish: both the result and the evasion of all sins.

To set the record straight, the "Brazilian" disinhibited sexuality is both an artifice and a farce. In fact, Antonio Herculano Lopes proposes that any "truth" related to it hides the ideological construction created around it (2003, p. 1). He suggests that the exaltation of Brazil as a privileged locale for the expansion of "free" libido was the result of a long and multi-layered process of symbolic construction that the Euro-Brazilian intelligentsia had to develop and put into practice (2003, p. 2). Certainly, the revistas' strategic repackaging of (old) Afro-Brazilian market women as sexy dancing mulatas was not an isolated case. Examples of tantalizing mixed-race figures that reproduce the ambiguous and hyper-sexualized baiana-mulata cluster pop up in theater as well as literature, visual arts, music, and fashion design. In a related article, however, Lopes (2001) points out that within Rio de Janeiro's revistas, the Brazilian intelligentsia finally found a secure and controlled space to (collectively) work out the ghosts and demons that they had inherited from their colonial past. In short, playwrights such as the author of *The Republic*, the prolific Arthur Azevedo, sought to construct humorous yet celebratory narratives about Rio/Brazil that could help that society to overcome or sublimate its atrocious past. Revistas portrayed the country as a euphoric community. Any guilt or pain that resulted from the unsolved slavery-miscegenation paradox was covered up or dissimulated by idealized sexy baiana-mulata or *mulatinhas sestrosas*. Hence, in reality, the revistas' celebration of passive and auto-exoticized objects of sexual desire obscured the historical violence of colonization, the inhumanity of the slave trade system, and the sense of shame regarding both sexual and cultural miscegenations. In the end, the popularity of these sexy dancing mulatas dressed as market women offers clear proof that Rio de Janeiro's patriarchal society was ready to incorporate Africa into its collective memory as long as it was presented as cheap and harmless objects for ephemeral consumption. Moreover, without black-face, this racially ambiguous character anticipates Freyre's "poultice" (i.e. mulatismo) addressed in the Introduction.

By the 1920s, multiple images of the baiana figure "populated" Rio de Janeiro's collective imagination. Beyond the carnival celebrations and burlesque musicals, images of "real" and "fake" baianas were mass-disseminated through posters of performances, photographs, illustrations, paintings, entertainment magazines, songs, books, fashion trends, etc. This figure also drew the attention of intellectuals, such as the poet, artist, and folklorist Cecilia Meireles. In her illustrated ethnography of Rio's "batuques," Meireles (1983) notes that, during the 1930s, carnival revelers from different ethnic and class backgrounds dressed as fake baianas. They continued to invent new trends such as shorter skirts, colorful shawls tied to the waist, and elaborate turbans with a little basket of paper fruits and flowers permanently attached to it. Instead of "real" religious amulets and jewelry, fake baianas wore an obscene amount of necklaces and bracelets made with oversized glass beads to the point that their heavy weight imposed limits on their mobility. Many also wore high-heeled slippers one size too small, which purposely made their walk leaping and unstable.

In the 1930s, the carnival revistas began to lose their audience to *chanchada* films, local musical comedies with equally tantalizing acts. Meanwhile, Rio de Janeiro's newspapers decided to promote these local black carnival parades in order to increase their readership outside the major sport seasons. In less than a decade, these black street parades and, by extension, the figure of the baiana-pastora became an unassailable source of local pride.[15] This rapid yet complex process started to take shape in 1929, when the locals of Cidade Nova organized the first samba parade competition at Praça Onze Square. Seeking to increase their sales outside sport tournaments, in 1932 *Mundo Esportivo* organized an official prize competition and repackaged these ex-Epiphany parades as exciting samba school parades. In the next few years, other local newspapers intensified their financial support and publicity to these once-marginalized processions, which was eventually matched by localgovernment aid. As part of these tournaments' regulations, all samba schools were obliged to: (a) include the "traditional" isle of baianas (*ala das baianas*) in every parade; and (b) exclude wind instruments from their syncopated marching bands. In 1935, Rio's city hall included these samba competitions in its official schedule of carnival activities. In the same year, the samba school *Portela,* then called *Deixa Falar,* won the contest with the first mechanical float in the history of samba parades: a baiana figure on top of a rotating globe. After 1937, a new policy required all samba schools to portray national themes. Under the Estado Novo's newfound sense of patriotism, the musical genre accompanying at the black parades, then called "hillside samba," was recognized as the official rhythm of Rio de Janeiro's carnival. Though conceived at hillside shantytowns, these baianas marched to the sound of a fast-paced samba across Rio's Parisian boulevards under tight regulations. These cumulative initiatives positioned fake baianas spinning their puffy colonial skirts at the center of patriotic-carnivalesque spectacles. Over time, the controlling regulation and "hygienization" of these prized competitions contributed towards further limiting the baiana figure's power to improvise ideas and to negotiate inter-subjectivity.

To summarize, the multiplying representations of the baiana figure that emerged in modern Rio de Janeiro, from street carnivals to staged musicals, had the potential to either expand or inhibit the choreographic power of negotiation of this black female personage. Because she was perceived as an "auto-exotic" figure, or perhaps despite that, "the baiana" became a popular carnival costume worn by men and women of different ethnicities and social classes, with different intentions in mind. During the first decades of the twentieth century, people forged several imperfect copies of the baiana figure, from white men cross-dressing as quitandeiras to black women as beauty queens, denting the previously established mode of signification with fractures and seams. The limitless originality of each individual characterization and the fakery of their multiplying street representations were formed by old and new stereotypes of the baiana figure. Beyond that, the juxtaposition between the unstable positionality of a marked character type and the tangible physicality of dissimilar bodies conjured up an unscripted semiotic space, producing ephemeral glitches. Unlike "real" ambulatory food vendors, for example, black carnival baianas were unbounded copies. Their performative glitches digressed from both the "alterity" with which subaltern bodies had been shamefully inscribed into scenes of subjection as well as the self-identifying "authenticity" which market women, Candomblé priestesses, samba party organizers, black matrons, etc. had proudly cultivated through their choreographies of self-identification. Finally, in 1939, Carmen Miranda exploded internationally in that particular role. As I will discuss next, the diva of Rio de Janeiro's showbiz generates, through her live and mass-mediated performances, a new fixed point of view from which to project Brazil as a modern nation-state.

Miranda: the syncopating queen within the Kingdom of Samba

When Carmen Miranda appears for the first time dressed as a "typical baiana" in a musical scene of the motion picture *Banana da Terra* (1939), her mass-mediated performance constitutes, as this chapter has sought to uncover, a bricolage of popular domain references. Amongst others, her characterization builds upon: "real" ambulatory vendors selling African-inspired food on the streets; Euro-Brazilian men cross-dressing as quitandeiras during Mardi Gras; colonial "naughty shepherdesses" shaking their hips in exchange for church donations during Epiphany; white-looking sopranos mimicking sexy dancing mulatas at urban stages; young women honoring Cidade Nova's tias baianas at black pageant-parades; high-society women dressed as "stylized" baianas at carnival balls, etc. As previously stated, the hyper-visibility of Miranda's choreographed enactment contributes towards the gelling together of samba, carnival, and the baiana figure into a fixed idea. Her immortalized persona combines, in particular, the desiring-but-not-honorable baiana-mulata type from vaudeville with the crystallization of its patriotic-carnivalesque counterpart from samba parades. In doing so, her choreographed performance fuses the racial indefiniteness and the tantalizing mockery characteristic of burlesque musicals with the dignified boasting and the hyperbolic fakery exercised at black carnival, thus collapsing the baiana-mulata-pastora who dances samba into a symbol of auto-exotic-yet-patriotic gaiety. That, my reader, is an acrobatic idea. In the end, though, the (auto-)exoticism of Miranda's character, especially her racial ambiguity and her desirable-but-not-honoring sexuality, perpetuates rather than resolves Brazil's pride-and-shame conundrum.

Released during the 1939 carnival, Miranda's interpretation of "O que É que a Baiana Tem?" for the camera should be understood in light of the socio-political effervescence of her times. Like her predecessors at the revistas, Miranda's white-looking celebration of choreographed blackness worked towards indirectly endorsing the elites' desire to whiten its population. Her acclaimed impersonation of the baiana figure on film was later re-enacted in live performances at the Urca casino, where it further catalyzed the already mentioned set of efforts that were being made towards the commodification and mass-dissemination of samba as a made-in-Brazil commodity. As previously mentioned, Miranda's samba-dancing baiana, a set choreography inside a glamorized scenario (casino show/cinema), illustrates, perhaps more than any artistic production of that time, the Estado Novo's "democracy of races" ideology.

The juxtaposition of Miranda's fair skin and her flawless diction against her soft-spoken hip-swings and her auspicious sense of humor provokes, more importantly, an epistemic shift within that nation's long-lived association across blackness, the ginga aesthetic, and shame. Mixing Euro-Brazilian and Afro-Brazilian cultural elements, her exuberant performance of this uncanny amalgamation further detaches cultural miscegenation from Brazil's violent – and violating – process of patriarchal colonization. Finally, Miranda's subsequent appearances in Hollywood motion pictures and Broadway musicals contributed to a dramatic change in perception regarding the value of Afro-Brazilian culture within the constructed image of Brazil. As Livio Sansone (2004) points out, Miranda's iconic visibility abroad worked as a seal of approval, jolting the judgment of taste in Brazilian society. Meanwhile, her unprecedented and unmatched stardom during the 1940s and the 1950s turned the *gingado* into an internationally famous way of moving that indicated *Brazilianess* beyond phenotypes.

It is possible that, at first, Miranda sought to either validate (give cultural recognition to) or valorize (add market value to) Afro-Brazilian culture without blackface with an exuberant rendition of a song about the hip-shaking abilities of black women. Nevertheless, as Miranda's international career progressed, the syncopated dialogues articulated between the hips and feet – so central to Afro-Brazilian dancing endeavors – were partially transferred to her ingenious coordination across curling hands, rolling eyes, and staccato moving tongue. Her hips and thighs were trapped inside fishtail skirts, her feet were strapped to platform shoes, and her head was layered with large turbans with artificial fruit and flowers. Still, artificiality does not easily define the efficiency with which her fake and indefinite choreography affected the Brazilians' sense of selfhood early on. The unbearable lightness of her syncopated polka-rolling tongue, her exquisite hand-eye coordination, as well as the precision with which she let herself go on stage, often making fun of herself, conjured up a hypnotizing grace and auspiciousness beyond words. In sum, Miranda transformed the impersonation of the baiana figure into an art form. Her meteoric explosion on global stages subsequently set in motion a new mode of identification that elevated indefiniteness and fakery into the official rubric of national(ized) authenticity.

Despite her financial and artistic accomplishments, Miranda was unable to avoid, or perhaps overcome, the crystallization of her artistic creation. In the end, the ambiguous character that once turned her into an overnight bombshell sensation, provoking a modernizing makeover of Brazil's international image, rapidly became a self-professed trap. In Hollywood musicals such as Irving Cummings' *That Night in Rio* (1941) and Busby Berkeley's *The Gang's All Here* (1943), Miranda's (fishtail) dances were no longer an auto-exotic representation of blackness that vivified the "democracy of races" ideology cultivated during the Estado Novo's populist dictatorship. There, Miranda's long-lived character stood for a new Latin American femininity stereotype endorsed by F.D. Roosevelt's "Good Neighbor" policy. Though the fake baiana acted as a floating signifier, in Brazil it had been historically associated with a particular gender, race, class, sexuality, occupation, etc. Nevertheless, on global stages, the mass-dissemination of a signifier grounded in fakery and indefiniteness (fake baiana) outside its contextual framework (Brazil's history of slavery and miscegenation) provoked a radical transformation in its meaning. In these Hollywood-ized narratives, the carnivalesque image of a samba-dancing market woman became an imperial metaphor for the ethno-cultural inferiority, hyper-sexuality, and docilized gender expected of all Latin American peoples.

Miranda's legacy, or rather the fixed metaphor this shooting star perfected, outlived her short career. By the time of her sudden death in the summer of 1955, other white-looking celebrities such as the ballerina and choreographer Eros Volusia (1914–2004; see Chapter 6) and the 1954 Miss Brazil, Martha Rocha,[16] had already contributed to further crystalizing the baiana figure as a static (or stale) symbol of Brazil's (auto)exoticness. Since then, mainstream initiatives in sectors such as entertainment, fashion, and tourism continued to exploit this floating signifier inside stereotypical narratives of/about Brazil. Though stylized images of baianas may evoke national or ethnic pride in contemporary contexts and discourses, in most cases such initiatives have further stripped the baiana figure of its power of negotiation.

In an article published in the *New York Times* entitled "Caricature and Conqueror, Pride and Shame" (1991),[17] the singer, writer, and composer Caetano Veloso outlines the conflicting effects that Miranda's impersonation of the baiana figure once evoked and continued to provoke in Brazil. In his opening lines, he states that:

> For a generation of Brazilians who reached adolescence in the late 1950s and became adult at the height of the Brazilian dictatorship and the international wave of counterculture – my generation – Carmen Miranda was first a cause for a mixture of pride and shame and later a symbol of the intellectual violence with which we wanted to face our reality, of the merciless gaze, we wanted to cast upon ourselves. (Veloso, 1991, p. H34)

As Veloso states above – and further unpacks in his autobiographical book *Tropical Truth* (2002) – the stereotypical representation of Brazil as a fruitful yet docilized samba-head left new generations of Brazilians with a bittersweet taste in their mouths. Veloso concludes that the "The Lady in the Tutti-Frutti Hat," "at once a disgrace and a deity" (1991, p. H41), had been the only representative of South America with universal legibility. This was exactly why, for him, the auto-parody became her inescapable prison. Above all, this Euro-Brazilian artist had, and continued to have, a defining impact within Brazil's mainstream sense of selfhood, so much so that by the late 1960s, the "merciless gaze" attitude of a group of artists led by Veloso and Gilberto Gil repositioned Miranda – both her invaluable contributions as an interpreter of Brazilian popular music and her canned impersonation of the baiana figure – at the center of their *Tropicalism* movement. By electing the *queen of camp* to illustrate their aesthetic and political concerns, both exalting her actions and turning her image inside-out, their counter-culture movement was able to expose and further question the pride and the shame surrounding modes of identity formation in Brazil.

Conclusion

In this chapter, I have sought to problematize the carnivalesque representation of baianas dancing samba as a fixed choreography of national identification. In my genealogy of the baiana figure, I first identified the relationship of black market women, or "real" baianas, to other Afro-Brazilian manifestations centered on the ginga aesthetic, such as samba and Candomblé, and how they employed this movement system in their everyday life working on the streets. Next, I unpacked two distinct, although interconnected, spheres where the baiana figure gained wider visibility in the aftermath of Brazil's abolition of slavery: teatro de revista musicals and samba school parades.

On proscenium stages, this marked figure was choreographed into fixed narratives of auto-exoticization associated with desired, yet not honorable, identities. Conversely, on the streets of Rio's carnival, Afro-Brazilian communities re-incorporated the baiana figure – both its attire and its syncopated moves – within their self-identifying discourses of resistance, connecting choreographed black motherhood to dignity and self-esteem. Contrary to the revistas and other scenarios where market women were represented with debauchery, at these black community-oriented events (the image of) this working-class female figure became a nourishing symbol of black pride, a point of departure from which to renegotiate visibility and dignity within the public sphere. In closing, I turned my investigation towards understanding the context under which Carmen Miranda's appropriation of the baiana figure was projected as a fixed symbol of Brazilian identity.

I arrive at the end of this chapter with the sense that it might be impossible to exhaust the reasons why the baiana figure came to occupy a central position within the modern conceptualization of Brazil as an imagined community. Instead, my genealogical excavation has led me to unearth the means through which Miranda's performances reiterated Brazil's pride-and-shame conundrum, the complex and unresolved set of ideas informing, and often distorting, the notion of Brazil as an imagined community. During Vargas' populist dictatorship, local cultural agents such as Miranda sought to orchestrate innovative projects aimed to celebrate the diversity of Brazilian culture in general and its African legacy in particular. Yet, while most projects celebrated Afro-Brazilian heritage, they seldom included black people in their productions. In addition, the reduction of the majority of Brazil's population to a few stereotypes, invariably represented by white-looking actors with or without blackface, contributed to restricting the pliability inherent in flexible choreographies of self-identification with which local blacks had negotiated their visibility, on their own terms, since colonial times. Hence, rather than solving this paradox, Miranda's white-looking mulata evaded rather than solved issues related to Brazil's slavery/miscegenation. Finally, on global stages, Miranda's stock character came to signify, or stand for, a "one-size-fits-all" model of Latin American identity.

Part III
Staging Ginga

6 Brazilian Bodies and Nationalism in Dance

At first, they were only a few; then, they were so many that this light source became a vision common to all. And then, they became a source of national pride, something one hadn't been in the habit of feeling for quite some time around here.

Helena Katz, *Grupo Corpo: Brazilian Dance Theatre*

In 1886, two years prior to the final abolition of slavery in Brazil, Rio de Janeiro's newspaper *Gazeta de Notícias* published *Terpsicore*, a short story by the mulatto writer Machado de Assis. There, the founder of Brazil's Academy of Literature (*Academia de Letras*) told the story of Porfílio, a working-class man who falls in love with a ballroom dancer named Gloria. Porfílio is enchanted, Machado de Assis tells us, by Gloria's way of dancing, not by her looks. This tropical version of the Greek and Roman muse of lyric poetry and dance (Terpsichore) moved, Machado de Assis writes, with a gracious and sensuous "mixture of swan and she-goat" (*mistura de cisne e de cabrita*: p. 20). Following the author's inclination towards irony, this metaphor choreographs the juxtaposition of unlikely ideas, capturing that which was too transgressive and indecent to be described, yet too intoxicating to be ignored. But what exactly is the meaning of this uncanny metaphor? What sort of imaginary interpolation could resonate with the offspring of a horned ruminant known for its lively and frisky behavior with a typically all-white water bird notorious for its phallic neck? And, furthermore, what kind of daring acrobatic feat would this "crossbreed" idea have to execute in order to transform this unsettling mixture into Terpsichore's "delight in dancing"?

Here are some clues. In nineteenth-century Rio Janeiro, "she-goat" was a slang term commonly associated with prostitutes and other women of "questionable reputation," often associated with black and mulatto women. Moreover, local newspaper readers would certainly know that Imperatriz, the street where Porfílio first sees his muse dancing in such a fashion, was the old Valongo Street. As the French painter Debret described in the 1830s, this is where "we find the black market, a true *trading post* where slaves are stored upon arriving from Africa" (emphasis added). Located near the docks, these so-called "Boutiques de la Rue du Val-Longo" were, as Debret outlines, "sometimes converted into a ballroom, with the consent of its owner." (1954, vol. 2, p. 78). Dancing in this case was a marketing strategy during slave auctions that was used to forge enslaved Africans as physically able (and thus profitable) bodies, despite incarceration, malnutrition, and the advent of the Middle Passage. In 1843, during the Imperial era, the arrival of the Emperor Pedro II's arranged bride, D. Tereza Cristina, at Valongo docks prompted a cosmetic beautification of that decadent corridor of greed, now renamed Imperatriz Street. With the end of the international slave trade in 1850 and the abolition of slavery in 1888 (signed off by the Imperatriz's daughter), these trading posts were gradually refashioned into import-export warehouses, rooming houses, and brothels. It was finally renamed Camerino Street (its current name) in 1890, when the country transitioned from a monarchy to a republican regime. Despite its name change, this historical street continued to function as a circum-Atlantic vortex of performance, interconnected by the trade of ideas, commodities, and human flesh, and punctuated by a racial hierarchy (Roach, 1996). There, at certain establishments of adult entertainment patronized by a wide range of individuals, from seamen to longshoremen, transcultural social dances such as the "inebriating" maxixe spread like wildfire (Efegê, 1974; see Chapter 3).

Therefore, it is safe to assume that while Machado's swan metaphor pushes his newspaper readers to stage, in their minds, the vision of auspicious prima ballerinas performing *Swan Lake*,[1] a dancing "she-goat" at the end of the nineteenth century in Rio de Janeiro pulls in the image of lascivious maxixe dancers, or *maxixeiras*, women whose dark skin tones and shameless ways of moving were often desired, but, as previously stated, were seldom valued or honored. What I find most intriguing about Machado de Assis' story is Porfílio's recognition of Gloria's dancing mixture as a graceful rather than shameful act. In fact, her nightlife discourse on dancing floors of "questionable reputation" her inebriating choreography of identification, leads her to marriage and, despite the financial constraints of her husband, to a life of social dignity and honor. In a sense, one of the greatest accomplishments of this postcolonial dancing muse is her eloquent ability to stir up aesthetic values

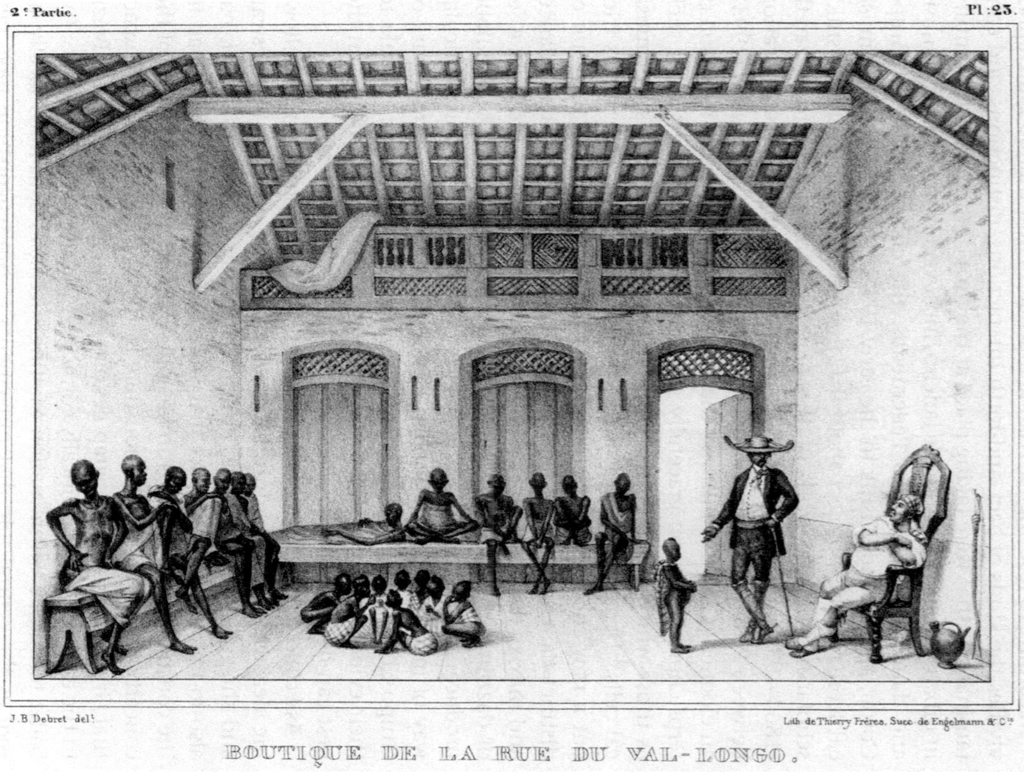

Figure 6.1 Jean Baptiste Debret's *Boutique de la Rue du Val-Longo*, 1839

associated with both the chaste and European-aspiring "swan-ness" and the local, muddy, and voluptuous "she-goat-ness." In doing so, the artful "hip(g)nosis" (Blanco Borelli, 2008) of her dance dismantles, or at least artfully blurs, the incompatibility imagined between these distinct ideas.

According to José Miguel Wisnik, Machado's now infamous metaphor vivifies the kinds of back-and-forth interactions, distinctions, and contaminations that were taking place at the turn of the twentieth century in urban centers such as Rio de Janeiro, most visibly in the areas of music and dance. When Rodrigo Pederneiras, the choreographer of Grupo Corpo, approached Wisnik in the early 1990s to commission the musical score for a new project about that unique moment in time, Wisnik's scholarship on the work of the piano composer Ernesto Nazareth (1863–1934) and Machado de Assis came in handy. The collaboration between Pederneiras and Wisnik resulted in the ballet *Nazareth* (1993), an eloquent "game of mirrors" that, like Machado de Assis' rendition of terpsichore, insists on gracefully rearranging contradictory ideas, such as those which Wisnik calls "erudite" and "popular" phrases or musical cells.[2] Departing from Wisnik's score, in *Nazareth*, Pederneiras tries out a new choreographic strategy that forges a refracting amalgam of distinct dancing technologies such as classical ballet, modern dance, and maxixe dancing. Pederneiras' innovative "poultice" produces a set of dazzling effect that, as I will demonstrate in this part of the book produces both desire and pride. A few years later, once this particular way of mixing local and foreign elements had become a trademark of Grupo Corpo, the dance scholar Helena Katz proposed that: "Thanks to their evident perfection, they wake up those who doubt that even the best can be 'made in Brazil'" (1995, p. 15).

As I collected these facts and opinions, I began to entertain the following idea: what if Gloria's intriguing way of dancing, her "mixture of swan and she-goat," holds the key to understanding not only *Nazareth*, but also to unlocking the impact that Grupo Corpo's repertoire has had in the construction of Brazil as an imagined community in the realm of the concert dance? As I will demonstrate in these two chapters, the impact that Grupo Corpo's choreographies had on contemporary dance in Brazil in the early 1990s is historically connected to the privileged status of the ballet technique within nation-exalting spectacles consumed locally, as well as the central role that movements of hip-syncopation have performed within exotic representations of Brazil destined for international audiences, from revue theater productions to folkloric ballets.

My hypothesis, which is the central focus of Part III of this book, is guided by a set of burning questions: what is the relationship between

Pederneiras' choreographies and Machado's acrobatic "mixture of swan and she-goat"? And, more importantly, is this way of mixing things implicated in the recognition of Grupo Corpo's repertoire as modern representations of the national in dance, which are both desirable and honoring? With that in mind, in these two chapters I will pay close attention to the ways in which Grupo Corpo mixes the corporeal structure of ballet with movements centered on swaying actions set in motion an innovative strategy for representing the particularities of Brazil's local culture in dance. Furthermore, I will investigate how and why this dance company's way of incorporating auto-exoticized elements such as hip-swings contributed to restoring Brazilians' sense of national pride in the 1990s.

"The national" in dance

My initial inquiries introduced a new set of problems. It begged the following question: what is Brazilian dance and how does Grupo Corpo's repertoire exemplify it? In a recent article regarding the construction of Brazilian identity during the 1970s, Daniela Reis proposes that "the obsessing preoccupation with representing *the national* has mobilized a good part of our artistic tradition and, in dance – a debate perhaps less known – this uneasiness is revealed from the first initiatives of construction and education of ballet in Brazil till more recent montages and researches" (2005, p. 2, emphasis added, author's translation). In this article, Reis qualifies her statement, pointing out that one should be careful not to make sweeping generalizations regarding the meaning of *Brazilianness* within concert dance. For her, "the national in dance has been represented in different ways, in distinct contexts and interests" (2005, p. 2). Along these lines, for Fernanda Peixoto (see Pereira, 2003), Brazil's re-occurring desire to stage "the national" on proscenium stages, as Reis puts it, should be understood as an intricate web of relations between hegemonic and non-hegemonic knowledges and between arts and politics.

In the case of Grupo Corpo, as I will demonstrate, its alleged "Brazilianness" lies in a series of interacting factors. Since its foundation in 1975, Grupo Corpo has sought to break away from the Eurocentric patterns of dance production established by state-sponsored companies. In addition, since the 1990s, the company has inaugurated a radical new era in which it has turned to prominently Brazilian themes and composers as a source of inspiration.[3] Consequently, Pederneiras has textured his choreographies with concepts specific to these local rhythmic palettes. Starting with the choreography *21* (1992), a watershed within the history of the company, one of the characteristics of Grupo Corpo's choreographic style has been its consistent articulation of a lexicon of movements distilled from a broad range of Brazilian dancing rhythms such as samba, maxixe, frevo, *xaxado*, and *forró*, already embedded in the musical scores. Yet, despite these interwoven textures, ballet (*balé*)[4] has occupied a central position in the company since its formative years.

In the documentary *Grupo Corpo 30 Anos: Uma Família Brasileira*, Carmen Purri, Grupo Corpo's former ballerina and current choreographic assistant, is categorical about this point: "All dancers of Grupo Corpo to this day need and must have a strong formation in classical ballet, which for us is the base and the departure point" (translated by the author). Hence, within Grupo Corpo's framework of representation, the auspiciousness of the *balé* technique – a form inherent from and representative of colonial power – provides a system of honoring support to the often explicit and overflowing sensuality of their hip-driven steps, gestures, and intentions that are characteristic of popular dance forms from Brazil. More importantly, Pederneiras has constructed a standardized syntax in which balé-trained bodies are pushed to interpret movement qualities inspired by Afro-Brazilian practices – and especially swaying movements – without losing or negating their balé foundation. As a result of Pederneiras' way of mixing things, Katz explains: "The mechanics of classical ballet, ended up being brewed into an entirely new medium. They were magnetized into daring angles from which sprouted all kinds of bumps and grinds" (1995, p. 23).

Similar to the deployment of the ginga aesthetic in the "Brazilian way" of playing football, Pederneiras' choreographies push dancers to expand their hips' range of motion towards "desencaixado" stances, thus "contaminating" the upward linearity of their corporeal core with decentralized juxtapositions, abrupt shifts, and other curvilinear dissonances. At the same time, he continues to choreograph from the point of view of balé. In other words, his formation as a ballet dancer in the 1960s and 1970s – his own *balleticized body* – continues to shape his choreographic choices, acting as an overarching force on his artistic creations, despite the local flavors he brings to the table. Beyond the aesthetic appeal of its choreographies, Grupo Corpo's particular way of mixing things has most importantly problematized the morality (e.g. the sense of pride or shame) of movements of hip-syncopation. In doing so, I propose, the company's interweaving of Africanist and Europeanist performative principles cultivated throughout the circum-Atlantic (Roach, 1996) has helped to dismantle the ontological battle over which aesthetic knowledges nationals from Brazil should subscribe to exclusively and/or hierarchic in their (self/othering) processes of identification.

Here it is important to clarify that although Pederneiras appropriates qualities of movements that are commonly associated with Afro-Brazilian heritage, Grupo Corpo's repertoire is not particularly invested in questioning the long-lived invisibility of blackness within Brazil's hegemonic context of artistic production.[5] Nor does it seek to create a literal translation, or transposition, of non-European practices cultivated across the country (e.g. samba school parades or Candomblé rituals) into the context of the concert dance (proscenium stage), as folkloric companies do.[6] Some of the company's dazzling performances have tilted dangerously towards exotic modes of signification. At the same time, the company's uplifting spectacles should not be simply interpreted as patriotic and over-optimistic portrayals of Brazil, imagined as a racially and culturally harmonious community, as famously theorized by Gilberto Freyre. Instead, within Pederneiras' choreographies, the sinuous and offbeat articulations of the hips have come to function as a frictional tool of mediation between the codified technique of (neo) classical balé and other sinuous, offbeat, get-down, or juxtaposed efforts permeating a wide range of regional movement practices. In doing so, this oscillatory juggling of ideas – or hip-play (*jogo de cintura*) – has occupied a central position within the asymmetric negotiations between two (or more) ways of articulating ideas corporeally within a single discourse. As I seek to demonstrate, Grupo Corpo's repertoire vivifies aspects of the frictional negotiations of power relations specific to Brazil's coloniality. In Pederneiras' choreographies, the strategic deployment of balé-trained bodies articulating a lexicon of movements centered on the ginga aesthetic reproduces a push-and-pull mechanism through which the concept of "Brazilianness" has acquired (a new/modern) legibility and credibility on both the global and local stages. "And then," Katz concedes, "they became a source of national pride, something one hadn't been in the habit of feeling for quite some time around here" (1995, p. 21).

One final issue requires our attention. The dancers at Grupo Corpo have interpreted Pederneiras' choreographies with such technical precision and efficiency that the transgressive juxtapositions they portray *seem* effortless. On local stages, for instance, Grupo Corpo's flawless code-switching between their foundational balé technique and other (local/street) movement vocabularies has contributed towards destabilizing, in the long run, the artificial incompatibility historically constructed between the Western way of dancing and "the Rest." Conversely, on global stages, the seamless ability of these balé-trained bodies to oscillate across a particular range of efforts associated with the ginga aesthetic is perceived as a "natural" or ontological, although exotic, aptitude attributed to or expected of Brazilians. Furthermore, because the company has consistently staged dances grounded in rhythms, images, and textures specific to the Brazilian ethno-cultural context, its repertoire is often mistaken for a representation of the exoticism/eroticism that is imagined to exist in Brazil. In other words, Grupo Corpo's oscillatory way of moving across hegemonic and non-hegemonic qualities of movement is at times regarded as a fixed set of determining and inevitable characteristics expected of (all) dancers born (or living) in that *exotic(ized)* locale. Internationally, Grupo Corpo's "mixture of swan and she-goat" (its balletic articulation of the ginga aesthetic) has become, ironically, a synonym of dance productions *from* and *about* Brazil. In Pederneiras' work, the indefiniteness generated by this playful juggling of ideas becomes, in and of itself, a symbol of Brazilianness in concert dance. In the already mentioned documentary that commemorates the company's thirtieth anniversary, for instance, Joseph Melillo, executive producer at the Brooklyn Academy of Music (BAM), defines Grupo Corpo as "precisely Brazilian" (Barreto and Santiago, 2007).

In sum, while the international recognition of Grupo Corpo's technical virtuosity and artistic excellence continues to validate and distinguish its name brand from its contemporaries in Brazil, on global stages, its members' oscillating swagger has been read (or *re*-cognized) as a "typical" example of Brazilian dance or, rather, a modern choreography of national identification. At the end, though, the constructed pliability and indefiniteness provoked by the flickering oscillations between undulating torsos and straight legs, jiggling hands and pointed toes, or turned-out positions and loose hips most often *replicate* rather than *solve* (or overcome) the epistemic frictions between "Euro-Brazilian" and "Afro-Brazilian" aesthetics, which are at the heart of Brazil's mode of identification: its pride-and-shame conundrum.

In order to better understand the impact that these multiplying interactions between balé-trained bodies and movement centered on the ginga aesthetic featured at Grupo Corpo's repertoire has had on the panorama of contemporary dance in Brazil since the early 1990s, I will first give a brief discussion of the historical role that the ballet technique came to occupy across the twentieth century within nation-exalting concerts catered to the local (European-aspiring) audience. In other words, I will outline the privileged position and the hierarchical power that the ballet technique has occupied in Brazil, contrary to the ginga aesthetic, as well as this embodied system of organization's ability to shape up local bodies under a Eurocentric episteme. My discussion departs from the already-addressed role that the ginga aesthetic assumed within the auto-exotic(ized) "made in Brazil" spectacle for export (e.g. revue musicals, casino shows, and carnival parades) since the end of the nineteenth century. Next, I will historicize the development of Grupo Corpo leading up to the production of *21* (1992) and *Nazareth* (1993).

In Chapter 7, I will offer a close analysis of *Nazareth* and *Breu* (2007). These two examples have been chosen based on the following criteria. First, each of these set choreographies is centered on post-colonial dance practices discussed in the previous chapters, which gained visibility at the end of the nineteenth century, when slavery was abolished in Brazil (1888). While *Nazareth* references maxixe, a syncopated ballroom rhythm closely connected to urban samba dancing in Rio de Janeiro, *Breu* makes references to frevo, a particular way of dancing historically associated with choreographies of violence that capoeira players improvised during street carnival parades in the twin cities of Recife and Olinda, in the state of Pernambuco. Secondly, there is an inverted parallel between the company's appropriations of blackness into their discursive metaphors. In *Nazareth*, ginga is auto-exoticized and the choreography celebrates hip-triggered sensuality inside a white/elitist environment. In *Breu*, there is a virilization of ginga and the choreography glorifies hip-triggered violence within a dark/public environment. Thirdly, they both deal with the concept of "multiple mirroring" that, as will be further discussed in Chapter 7, off-centers dichotomies such as self/other or pride/shame into kaleidoscopic ambiguities, with three or more kinds of efforts. Finally, the 15-year gap between the two choreographies provides the necessary distance alongside which one can re-trace, measure, and qualify the increasing incorporation of the ginga aesthetic within their repertoire. In doing so, this study sheds light on the processes of reconstructions and deconstructions – or rather decolonizations and re-colonizations – exercised in Pederneiras' choreographic style as well as Grupo Corpo's performances.

Ballet, coloniality, and the cultivation of civility in concert dance in Brazil

In 1909, Mayor Pereira Passos concluded his hygienist gentrification – or "de-Africanization" – of downtown Rio de Janeiro with the inauguration of the city's Municipal Theater (*Theatro Municipal*). It is a modern opera house built essentially as a downsized replica of Paris' Opera *Palais Garnier,* France's national opera house in Paris. Built during the Brazilian Belle Époque (1889–1922), the inauguration of this theatrical venue symbolically marked a radical change in the life of Brazil's "postcard city" towards a new vision, whose motto was "Rio becomes civilized!" Subsequently, Rio's Municipal Theater established itself as an unprecedented trading post of international culture and civic enlightenment, where the local European-aspiring elites could go to import, experience, and reproduce Western aesthetic forms and values. Offered on its menu were Italian operas, Russian ballets, and French-style balls and masquerades, all interpolated by European languages, fashion, and etiquette. More importantly, this performing arts building enabled the inclusion of Rio de Janeiro within international expeditions, whereby European ballet companies circumnavigated the world, staging and disseminating their choreographed ideas about the delight in dancing (Paixão, 2009). In the following decades, a great effort was made to establish that replica of Paris' opera house as the heart of the "modern and civilized" culture of Rio de Janeiro. In 1927, it invited the Russian choreographer Maria Olenewa (1896–1965), a former prima ballerina with the Ana Pavlova Dance Company, to direct its *Escola de Danças Clássicas do Theatro Municipal do Rio de Janeiro*. In 1947, Vaslav Veltchek (1896–1967) was invited to direct a similar state-sponsored classical ballet school in São Paulo's Municipal Theater.

Contrary to the ginga aesthetic, which was perceived as a "shameful" and "corrupting" way of dancing, on local stages in Brazil the balé technique has exercised a "chaste" and "ennobling" role. Imported from Europe and praised at elitist venues such as Rio's Municipal Theater, the ballet technique seeks to discipline dancers to sustain a fixed, turned-out position of the pelvis. Since the hips are aligned with the torso of the body as a unit in a "locked-in" position (that in Brazil is referred to as "encaixado"),[7] balé's codified technique privileges movements instantiated from the legs and arms outwards.[8] Conversely, movement practices centered on the ginga aesthetic articulate pliable and dynamic movement of the hips independent from, or out of alignment with, the rest of the body (referred to in Brazil as "desencaixado").[9] In addition, the balé's historical pursuit of upward linearity has been associated with the chastity and honor expected of "decent" dancers. By contrast, the "sinuous," "offbeat," and "get down" accents associated with the ginga aesthetic, as staged in revue theater productions (teatro de revista; see Chapter 5), has been associated with non-white ethnicity, and a lack of sexual morality. Hence, beyond the formal differences between these two ways of articulating the body, the polar-opposite value or worthiness allocated to the balé and the ginga aesthetic or, perhaps, *swan-ness* and *goat-ness,* rearranges these distinct movement systems into exclusionary dichotomies. Within this genealogy of concert dance in Brazil, in particular, the ballet technique and the ginga aesthetic have been organized as polar-opposite technologies.

One of the crucial points supported at the Municipal Theater's school of dance and carried over throughout most of the twentieth century across the country was the belief that the ballet technique was the only way to produce civilized dancing bodies and, as discussed below, honorable narratives about the nation. In that sense, Katz affirms, in Brazil, the balé has functioned as a "benchmark of colonial power" (1995, p. 23). In practical terms, suffice it to say that the Municipal Theater, its mimetic architecture, its imposed dress code and etiquette, its highly regulated repertoire, its staged narratives, alongside its regimen of bodily discipline of its state-sponsored dance school and its exclusively white cast, all contributed to keep Brazil's coloniality alive. Under the rubric of progress and civility, this body politics installed a process of stratification of different systems of organization, production, and dissemination of embodied knowledges according to an exclusionary dichotomy. The Western way of dancing, for instance, was praised and safeguarded inside the walls of the Municipal Theater. The local, primitive, half-breed, or contaminating "Rest" remained "out there," beyond those walls.

During the 1930s and 1940s, Getúlio Vargas' paternalistic administration further invested in two major fronts in the area of performing arts. On the one hand, it sought to include "popular culture" and "folklore," especially those connected to the African legacy in Brazil, within populist discourses of national identity. In this scenario, the ginga aesthetic remained connected to that which was primitive, immoral, irrational, child-like, or linked to the colonial past. On the other hand, Vargas' administration began to sponsor (financially) and support (politically) concert dance productions centered on the balé. In 1936, for instance, Rio's Municipal Theater established Brazil's first *corps de ballet,* which followed the European system of hierarchic stardom and typecasting. State-sponsored ballet companies, such as the Municipal Theater's corps de ballet, focused on two distinct kinds of productions: (a) the re-staging of quintessential pieces from nineteenth-century and early twentieth-century European ballets (e.g. *Swan Lake* and *The Nutcracker*) that fulfilled a national demand for "culture and civility";

and (b) the creation of original "Brazilian ballets" (*Bailados Brasileiros*), which often catered to an international audience. The expression "Bailados Brasileiros," a concept first coined by the Brazilian dancer and choreographer Eros Volusia (1914–2004),[10] was later appropriated and refined by (mostly foreign) choreographers working in Brazil.

Generally speaking, these so-called "Brazilian Ballets" privileged "ethnic" narratives (e.g. Indigenism) shaped by a Eurocentric understanding of citizenry and a Romanticized notion of nationhood, tamed by Roman Catholic notions of morality. Considered an exemplary piece of this genre, Carlos Gomes' opera-ballet *O Guarani* (*Il Guarany*, 1870, originally libretto in Italian) tells the forbidden love-story between the daughter of a Portuguese nobleman and a warrior of the Guarani tribe. Based on José de Alencar's novel *O Guarany* (1857), a classic of Brazilian Romantism, Gomes' opera was first staged at Rio's Municipal Theater in 1937, at the dawn of Vargas' Estado Novo. Here, as in many other Amerindian "bailados," ethereal and white-looking ballerinas re-dressed as *beau savages* leap between *grand jettes* and *pirouettes* with an uncompromised gayety, against backdrops filled with tropical flora and fauna. Collectively, these early productions choreographed Brazilianness inside an auto-exoticized "lost paradise" discourse that, divorced from socio-political conflict, tended to privilege idealized encounters between Europeans and Natives, invariably erasing Africans and their descendants from Brazil's myth of origin. In doing so, these Romantic narratives contributed to further de-historicizing the physical and epistemic violence of the European process of colonization implemented in Brazil, both the genocide of Amerindians and the traffic in and enslavement, torture, and rape of African nations.

During these first decades, choreographers working for state-sponsored companies made rare attempts to stage bailados centered on Afro-Brazilian themes. In most cases, initiatives such as Olewana's *Maracatu do Chico-Rei* (1939, performed in blackface) were strongly rejected by the patrons of "High Art" venues such as Rio's Municipal Theater (Pereira, 2003). In the end, the exclusionary adoption of classical ballet (medium/language) and its nineteenth-century Romantic narratives (scripted discourse) at concert stage productions functioned as didactic tools with which to indoctrinate national citizens with a Western sense of aesthetic taste and character. In particular, the classical balé's disciplinary system of bodily organization grounded in the turn-out position and its codified pursuit of (Cartesian) precision, uprightness, linearity, and weightlessness as virtues were adopted as the only regimen capable of producing delight in dancing. The "immunological" control exercised within the protective walls of the Municipal Theater (and other venues like it) maintained, above all, hierarchic articulations of space and aesthetic strategies of dependency.

After the Second World War, Brazil's local avant-garde artists rebelled against the aesthetic autocracy of classical ballet, replacing it with other foreign influences such as modern dance, expressionist dance (*expressão corporal*),[11] or the neo-classical balé.[12] Meanwhile, choreographers such as the Afro-Brazilian Mercedes Baptista (b. 1921),[13] a student of Eros Volusia and Katherine Dunham (1909–2006), searched for new ways to transpose Afro-Brazilian dance practices onto the proscenium stage under the rubric of "folkloric ballet." Contrary to most modern dance choreographers' attempts to represent the "Brazilian reality" on the stage, Baptista's efforts gained wider visibility on global stages, indirectly reinforcing Brazil's imagined exoticism. Her efforts were followed by other similar companies such as *O Balé Popular de Recife*, which was founded in 1977.

During most of the 1960s and 1970s, a period that coincided with the height of Brazil's military dictatorship (1964–85), we see the emergence of overlapping counter-culture ventures put in motion by the intellectual left, such as *Grupo Oficina* (theatre), *Arena* (theatre), *Tropicalism* (integrated arts), *Cinema Novo* (cinema), and *Pasquim* (newspaper). Within the realm of concert dance, avant-garde choreographers such as the founders of Ballet Stagium, Décio Otero (b. 1951) and the Hungarian Marika Gidali (b. 1937), as well as Ruth Ranchou (b. 1927), Marilena Ansaldi (b. 1934), Célia Gouvêa (b. 1949), Klauss Vianna (1928–92), Angel Vianna (b. 1928), and the Argentinean Oscar Araiz (b. 1940), would collectively criticize the representations of the national in dance that was characteristic of the first part of the twentieth century. For Daniela Reis:

> If from the 1930s to 1950s the language of ballet revealed itself under the desire to conceive a "national bailado," wrapped in the internal logic of the European *ballet* (school, dance company, official theatre, hierarchy of roles) and, still, bringing characteristics of Romantism [to the stage], even if belated. Later on, those who adopted "modern dance" would contest this language, because they did not approve of the romantic and the European parameters of ballet. To the subscribers of the modern style, the Brazilian dance could not be expressed through such a codified and systematized technique whose tradition derived *from aristocratic courts*. The Brazilian modern dance, in full process of consolidation and construction of its history, would face

> issues that put into question the corporeal form of bodily technique of representing human beings and Brazilian issues. (Reis, 2005a, pp. 2–3, emphasis in original, translated by the author)

As Reis outlines above, during this period, "Brazilian dance" came to be understood as any spectacle that addressed local socio-cultural themes with a realistic – or perhaps regionalist – approach. More specificially, these "modern" choreographers opposed the adoption of Romantic narratives and the classical ballet technique as national standards in dance as well as the eroticism/exoticism with which the national dance had been choreographed for export. Companies such as Ballet Stagium, for example, incorporated both Amerindian and Afro-Brazilian themes into their choreographies. Gidali and Otero, the company founding directors, also made a conscious effort to stage their dances beyond the Rio–São Paulo circuit, touring with their company in both little towns and big cities across Brazil, and also performing inside public schools and in open parks. "The 70s theme was almost self-explanatory," Katz concludes, "to make dance was a way of positioning human beings in their environments" (1994, p. 76).[14] This cutting-edge "sub-class" of modern dancers invested in choreographies that, as Katz and Reis seem to agree, sought to situate or contextualize Brazilian citizens within their "real" ecologies or the particularity of an imagined local/regional "reality." Still, despite their significant contributions and the radical transformations that these artists set in motion, in the 1970s, most of these companies continued to employ (neo-classical or modern) ballet as the primary regiment of training, even when they performed "expressive" or "contemporary" dance on stage. This was also the case for Grupo Corpo, whose debuting choreography *Maria, Maria* (1976) portrayed the harsh reality of life for working-class women in Brazil, but whose dancing bodies still evoked upward linearity, frailty, and weightlessness. As Reis concludes, this viewpoint was supported in Brazil by the "belief that ballet technique is better able to prepare the body of a dancer, for it is 'cleaner', 'complete,' and that, if a dancer becomes familiar with *balé*, he is capable of dancing any style" (2005a, pp. 7–8).

Overview of Grupo Corpo's development from the 1970s to the 1990s

Grupo Corpo's journey began in 1971 when the Pederneiras siblings and a group of their friends got together in Belo Horizonte to explore choreographic alternatives to the typical national dance forms. During this period, the Pederneiras siblings even convinced their parents to move out of their family home so that they could use it as a dance space. Among its initiatives, in 1973 the group attended Ouro Preto's Winter Festival of Dance, where the members met the Argentinean choreographer Oscar Arraiz. At that event, the choreographer offered this group of dancers an opportunity to study classical ballet in Buenos Aires. Subsequently, in 1975, the Pederneiras siblings (four brothers and two sisters), along with six friends, founded Grupo Corpo. Under the artistic direction of Paulo Pederneiras, the company hired Arraiz that same year to choreograph Grupo Corpo's debut work, *Maria, Maria* (1976). It also commissioned Fernando Brandt to write the script and Milton Nascimento to compose the original score. *Maria, Maria* was a danced theatrical performance whose theme centered on the specificity of Brazilian reality. It was an instant success and traveled to 14 countries over the span of six years. Following this knockout tour, the emerging company invited the same group of artists to produce *Último trem* (1980), a concert dance piece that also addressed Brazilian reality with the same dance-theater approach.

Strongly influenced by Arraiz, in 1978, Rodrigo Pederneiras made his debut as a choreographer with the work *Cantares* and, in 1981, he assumed the role of Grupo Corpo's resident choreographer. In that same year, Emílio Kalil became informally involved with Grupo Corpo, officially sharing the position of co-director from 1987 to 1989, when he was invited to direct Rio de Janeiro's Municipal Theater. In 1981, the company also opened its new (and still current) location: a three-story facility with dance classrooms, an art gallery, and a 200-seat theater, where the company takes classes and rehearses daily. During this phase, Grupo Corpo produced a series of works choreographed to Eurocentric classical music drawing from both foreign and Brazilian composers and systematically grounded in neo-classical ballet technique. With *Prelúdios* (1985), first performed at the First International Dance Festival of Rio de Janeiro, Rodrigo Pederneiras gained national recognition as a professional choreographer. *Prelúdios* was the group's first collaboration with Freusa Zechmeister, who remains the company's costume designer to this day.

In 1989, Grupo Corpo signed a renewable three-year contract with Shell Brasil that would eventually last a full decade, during which time Shell remained Grupo Corpo's primary sponsor. Grupo Corpo was the first dance company in Brazil to obtain maintenance sponsorship from a large corporation and, despite the company's national status, the budget increase enabled it to finance performances that had an "almost operatic dimension, in the sense of a close collaboration amongst the arts" (Bogéa, 2001, p. 135). The partnership with Shell allowed the company to increase its national and international tours and its staff. According to Inês Bogéa, a former dancer at Grupo Corpo: "The partnership, which lasted from 1989 to 1999, also pushed toward worldwide recognition for Corpo, which began to travel throughout Europe and the Americas on a routine basis" (Bogéa, 2001, p. 134). In addition, Grupo Corpo's financial stability had an enormous impact on the quality of its professionals. In the 1990s, for example, it was one of the few dance companies outside the classical (state-sponsored) circuit to offer professional stability to dancers through competitive salaries and health insurance, frequent performances, and international exposure. Thus, in the 1990s, it became the "dream" of many talented artists seeking alternatives inside Brazil. Many Grupo Corpo dancers have migrated, upon their retirement from their stage career, to other functions in the company, from administrative to technical positions. As a consequence of its financial stability, the company was also able to extend its international standards of technical excellence across local stages, while touring outside the Rio–São Paulo circuit. Beyond its artistic core, the Shell partnership enabled the company to expand and diversify its technical staff in terms of both quantity and quality. Since 2000, under the sponsorship of Petrobrás, this financial stability has translated into a 60-employee corporation, with a permanent staff of dancers, ballet teachers, choreographic assistants, stage technicians, and administrative personnel. Instead of relying on what minor theatrical venues have to offer, Grupo Corpo travels with its own technical equipment and trained staff, which may range from stage lights and offstage technicians to a full-size (movable) stage. Finally, the increase in the size and frequency of international tours has contributed to the recognition of Grupo Corpo as a national brand name on global stages.

From 1989 to 1992, Grupo Corpo's newly consolidated creative team led a series of gradual transformations that culminated in the ballet *21* (1992), which was mentioned earlier. First, in *Missa do Orfanato* (1989), Rodrigo began to break away from the linearity of neo-classical vocabulary characteristic of his previous phase and delved into an investigation of the Brazilian Baroque. Following *Missa do Orfanato*, which was choreographed to Wolfgang Amadeus Mozart's *Weisenhaus Mass*, Grupo Corpo's creative team produced *A Criação* (1990), choreographed to Franz Joseph Haydn's *Creation*; *Variações Enigma* (1991), choreographed to Edward Elgar's *Enigma Variations*; and *Três Concertos* (1991), choreographed to three of Georg Philipp Telemann's *Concertos in D*. Although the works are embedded in classical music, in each of them Pederneiras progressively incorporated movement aesthetics found locally in Brazil in order to twist both the linearity and the weightlessness of classical ballet. Across these four works, Pederneiras took the liberty of stretching or bending, here and there, the codified technique of neo-classical or modern balé, giving a unique local flavor to his choreographies.

According to Bogéa, in *A Criação*: "Rodrigo's choreography uses classical ballet as a point of departure and transforms it into other material: he gradually incorporates angles, sways, and swings, creating a new grammar of movements. This new mixture of aesthetics which was already visible in *Missa*, now becomes, more than just visible, thematic" (2001, p. 135). But this was a gradual process of transformation. According to Helena Katz:

> *Prelúdios* (1985) changes everything. Rodrigo is beginning to be considered one of the top professional choreographers and the whole country begins to celebrate him as an artist to watch closely. From *Prelúdios* (1985) to *21* (1992), thirteen choreographies are produced, in which he performs a series of experiments that dominate progressively more structures and spaces, leading the classical technique of ballet to become apt to receive transforming contaminations. (2000, p. 313)

It should be noted that within Grupo Corpo's first ballets in the 1990s, these warping transformations were visible on the vocabulary level, but not in terms of syntax. Hence, despite these "transforming contaminations," the overall aesthetics of Pederneiras' work continued to reflect Eurocentric episteme, from the dancer's strict system of bodily organization (classical dance) to the themes and musical scores (classical music).

The turning point

As mentioned above, the choreography for *21* (1992) inaugurated a radical new phase in the trajectory of Grupo Corpo because it marked the company's turn towards (mostly) Brazilian musical rhythms and themes as a source of inspiration. The choreography was a collaboration between the company's creative team and the musicians of the percussion ensemble Uakti. Considered a watershed in the history of the company and a benchmark within the genealogy of concert dance in Brazil, *21* introduced new ways to articulate "national" ideas in concert dance. Thus, it complicated what it means to be Brazilian within the mainstream sphere.

Choreographed to the unique sound of Uakti, *21* replicates the mathematical polyphony available in the original musical score, and its irregular meter signature, in two distinct ways. During the first half of the score, Pederneiras deconstructs the composite number 21 into its constituting prime number (7, 5, 3, 2, and 1). This organizational matrix is transferred onto the stage with mechanical-minimalist sequences of multi-meter gestures and steps, making the structure of the music visible three-dimensionally. This structural organization influences the rhythmic organization and distribution of bodies in space (i.e. solos, duets, and groups of 3, 5, and 7 dancers) as well as the frequency in which movements are repeated or arranged. These precise sequences of movements are tied to the musical score in such a hermetic way that the dancers on stage evoke not only the melody but also the notes themselves, moving across a hypothetical sheet of music. In the second half of the ballet, the choreography radically switches to bold and festive movements centered on bodily swing. Synchronized with the oversaturated lighting and the bright-colored costumes and set design, the dancers' sinuous articulation of their hip joints impregnates the stage with references to Brazilian regional images and symbols. Contrary to the synchronized minimalism of the first half, in the second half one notices subtle variations in the dancers' interpretation of Rodrigo's choreography based on the embodied knowledge that each interpreter carries within his or her muscle memory.

Despite the colorful ebullience and clear references to samba and carnival, *21* should not be dismissed as a display of alienating patriotism. In order to fully understand the impact that *21* had nationally in the early 1990s, it is important to also take into consideration the historical context in which both the company and the audience were immersed. Politically and culturally, 1992 was a critical year in Brazil. Among the events that marked the country profoundly, it is worth mentioning the student movement known as "painted faces" (*caras pintadas*), which demanded the impeachment of President Fernando Collor de Mello. As part of this youth-driven political movement, students took to the streets during August 1992 to demand the President's impeachment due to accusations of corruption. The term "painted faces" refers to the form of visual expression that became the symbol of this patriotic manifestation: students painted their faces in the colors of the Brazilian flag. At the same time, the musical scene witnessed the international explosion of the samba-reggae band Olodum, an Afro-Brazilian cultural group based in Salvador and founded by the percussionist Neguinho do Samba. Created in 1979, its original goals were to combat racism, to encourage self-esteem and pride among Afro-Brazilians, and to fight for civil rights. Neguinho do Samba's Olodum is credited with introducing a unique rhythm for the 1986 carnival, later known as samba-reggae, which mixed samba with reggae, merengue, and calypso. Olodum's performance on Paul Simon's 1990 album *The Rhythm of the Saints* and at the singer's subsequent concert in New York City's Central Park in the summer of 1991 established the group's international fame. Later, on October 2, 1992, an unprecedented massacre took place at Carandiru Penitentiary in São Paulo. Considered a major human rights violation, the massacre was triggered by a prisoner revolt and resulted in the killing of 111 inmates, 102 by gunshots fired by the military police. None of the 68 police officers was killed. So far, no one has been convicted in relation to this major incident.

Visually and kinesthetically, the use of balé-trained dancers interpreting popular street culture in *21*, along with the technicolor celebration of local flavors and textures, resonated with the wave of unsettling imaginations of Brazilianness that surfaced at the time. First, while street demonstrations were a common practice at the time, the public expression of national pride of the young "caras pintadas" was unprecedented. Second, the international explosion of Olodum's exquisite rhythms and innovative choreographies of self-identification, in which blackness is spotlighted as a code of honor, caused many Brazilian citizens to feel (and express) pride in their African heritage. At the same time, these evocative moments were unable to overshadow the erupting shame that events such as the accusations of corruption under Collor de Mello's administration or the police brutality at Carandiru provoked within this re-emerging Brazilian spirit. Although festive at first glance, given the specificity of this socio-political context, the negotiations involving the ginga aesthetic in *21* had the power to cast new light on the same set of transferred (and unresolved) ideas available in Brazil since colonial times, its pride-and-shame conundrum.

Starting with *21*, the company moved in a completely new direction, exploring renewed combinations between the ballet technique and local non-European textures, rhythms, and hues. In the following year, it presented *Nazareth* (1993), choreographed to an original musical score by José Miguel Wisnik and based on the work of the composer Ernesto Nazareth. This piece represented the company's first attempt since the inaugural *Maria, Maria* (1976) to foreground both local artists and local(ized) ideas (texts, rhythms, movements, and images) in order to construct a spectacle addressing the specificity of the Brazilian cultural context. More importantly, this was the first time that the balé-trained dancers of Grupo Corpo employed bodily syncopation across the entire choreography. In *Nazareth*, synchronized swaying of the hips enveloped in playful and voluptuous "ragtime" rhythms functioned as choreographic strategies to portray desire.

Dancing between local and global stages

Since its creation, Grupo Corpo has staged more than 20 original productions, of which about ten remain as part of its permanent repertoire. Over the years, it has reinvented itself a few times, while maintaining a positive relationship with a captivated audience and the specialized press. This recognition extends both to the artistic consistency of its staged performances – the creative exuberance of its spectacles and the technical competence of its onstage and offstage professionals – and to the financial stability and productivity of the company's organizational structure, which includes its physical facilities, salaries, marketing strategies, and corporate culture. Across the globe, it is known for its dazzling spectacles, which fascinate audiences with what is perceived internationally as a distinctively "Brazilian" way of moving.

On local stages, the work of Grupo Corpo has provoked a radical semantic shift in the way in which the ginga aesthetic is incorporated into proscenium stage productions, which surpasses previous attempts made by some of its predecessors, especially the iconic Carmen Miranda, Eros Volusia, and Mercedes Baptista. In particular, it has constructed a personalized way of dancing within high-art venues, in which bodily syncopation becomes legible within the choreographer's own terms and is not translated as primitive or erotic dance. In contrast to the less honorable message that hip-shaking typically projected at populist genres such as the revue theater, casino shows, and samba school parades, by juxtaposing offbeat swings and balé, Grupo Corpo has produced a signature style that simultaneously generate desire and pride nationally. Its dancers' balleticized bodies provide an honoring structure of support over which the ginga aesthetic "makes sense" and thus make the ticket-buying local elites proud of *identifying with* what they see on stage. In fact, some of Grupo Corpo's most passionate followers in Brazil – from art critics to local fans – maintain their assiduous devotion to the company because its spectacles foreground the specificity of Brazilian cultural themes.

On global stages, a fair number of intellectuals and dance enthusiasts have commented on Grupo Corpo's work. Yet, whenever the connection between the company's non-verbal form and foreign content exceeds the cultural references of the audience, the performances generate a two-pronged reaction in which the choreographed discourses are applauded for their entertainment value and at the same time are labeled as virtuoso spectacles that portray exotic otherness. For instance, some of the reviews of Grupo Corpo's performances in the *New York Times* emphasize the vacuum of multicultural darkness between the proscenium and the audience that not everyone is able to traverse (Rockwell, 2005; Sulcas, 2008). Grupo Corpo's choreographies are fast-paced enough that foreign spectators can appreciate the dance without being fully aware of the cultural context and the implied meaning of what they see. From this point of view, Rodrigo Pederneiras' mixed-scores, combined with the technical ambidexterity of Grupo Corpo's dancing bodies and the overall aesthetic pleasing look that Paulo Pederneiras brings to their spectacles, have the mesmerizing effect of a hypnotic opiate.

In an interview given to the *New York Times* in 1996, for instance, the director of the Lyon's Biennial Dance Festival, Mr Darmet, stated that, in general: "What fascinated me with Brazil is that the relationship with the body is so *natural*." Referencing Grupo Corpo, he concluded that "Brazilian dancers, even the most avant-garde, are very tied to their *roots*, to *nature*, to the *sea*, to the northeastern *outback*. They're more expressionist than what we're used to in Europe" (Riding, 1996, emphasis added). Despite the virtuoso expertise and level of excellence that the company has developed over the years through disciplinary training, when performed outside Brazil, the audience seems particularly fond of the "naturality" with which dancers from Brazil are in contact with their "primitive" or "outback" side. More often than not, the company is admired internationally largely for its trademark accents of exotic-erotic hip-swings and endless physical vitality and fitness rather than for its display of ballet-like virtuosity or its intellectual creativity which it is renowned for in Brazil. Although some critics and enthusiasts are able to recognize the subtle appropriations and deconstructions vivified on stage, in general these international audiences take the balletic body for granted, while noticing how its movement vocabulary differs from the (expected) norm. At last, despite the company's commercial success (or perhaps because of it), at times its spectacles have contributed to reinforcing the exoticism imagined about Brazil.

In recent times, some Brazilian dance scholars have problematized Grupo Corpo's rhetorical strategies as well as the context in which its work has been framed internationally. According to dance scholar Paulo Paixão, the work of Pederneiras continues to reiterate all the old stereotypes associated with Brazilianness such as "hip shaking" (*requebros*), "smooth undulations" (*moleza*), and "tenderness" (*meiguice*) (Paixão 2009, pp. 99–101). For Paixão, this stereotypical depiction reiterates the image of a utopian happiness attainable through partying, dancing, and sex, in the midst and in spite of the poverty, the quotidian violence, and the striking socio-economic contrasts between the rich and poor. Furthermore, he points out that events such as Lyon's 1996 Biennial Dance Festival, through which Grupo Corpo became widely known in Europe, have functioned to validate dance productions from marginalized parts of the globe and to introduce these new products to the international/European market. In addition:

> This panoramic way of exhibiting dances relates to a differentiated context of circulation of information and the reduction of the complexity of the local phenomena motivated by a so-called globalized democratization of singularized cultures. Along with the radicalization of capital's globalization, the cultural diversity stereotype is a way to camouflage the unequal profit between the different countries. (Paixão, 2009, p. 94)

Under the rubric of "cultural diversity," these global stage festivals have contributed to reducing the international audience's understanding of Brazil's artistic production to just a handful of cultural goods. In the realm of concert dance, companies such as Grupo Corpo and Deborah Colker have been selected to represent Brazilianness on the world map of national dancing.

Yet, I disagree with Paixão's view of Grupo Corpo's choreographies as stereotypical visions of Brazil, that is, as a pleasure-seeking or alienating blend of cultural traditions. In each spectacle, the company's rhetoric continues to unwrap and stack up on stage eclectic units and mix-matched fragments extracted from different ethno-cultural segments. When juxtaposed on stage, these coexisting epistemes provoke surprising effects that ultimately make local(ized) aesthetic principles intelligible to a broader audience. In addition, these formalist oscillations offer a new way to approach the question of what it means to be Brazilian. Rather than presenting the audience with spectacles symbolizing the cultural homogeneity of Brazil, the singularity of Grupo Corpo lies in the choreographer's artful ability to string together distinct kinds of efforts and movement qualities locally practiced in Brazil into set choreographies. Often, this includes dialoguing with the heterogeneous strains of aesthetic knowledges already embedded in the muscle memory of its dancers. Nevertheless, on stage, the subtle apartness of each dancer is counterpoised by his or her shared subscription to the balé technique as a foundational regimen of training.

Grupo Corpo's dancers move in and out of sync with erudite/Eurocentric and street/Afrocentric embodied technologies. In doing so, its choreographies portray Brazilianness as a flickering and ever-changing process filled with affirmations and negations. As I will demonstrate in the next chapter, Pederneiras' rhetorical strategy takes advantage of the interpreters' ambidexterity or polyglotism to represent the friction historically constructed between Western systems of organization and production and "the Rest." At the same time, he off-centers fixed oppositions such as "turn-out position" and "hip-swing" with multiple (three or more) efforts.

Collectively, Grupo Corpo's staged choreographies articulate a flexible system of non-verbal speech patterns that questions the artificiality through which embodied knowledges have been recognized as a determining factor within ethnic or national processes of identification. This strategy aligns the company's work with other avant-garde movements that have emerged in Brazil, which have placed "popular" and "erudite" (or Western and non-Western) spheres in dialogue with one another, thus blurring the lines between them. At the same time, it is necessary to take into consideration the specificity in which these moving bodies are situated and the point of view from which they speak: the ballet technique. Also, its production cannot be fully disassociated from the complex processes of auto-exotification exercised within local proscenium stages since the nineteenth century in the revue theater and other popular venues, and their multiple repercussions on global stages. In the case of Grupo Corpo, the Euro-Brazilian balé has unquestionably worked to "clean up" and "moralize" the dancers' hip thrusts, while their characteristic hip-swinging gives concert dance an irresistible sex appeal. Above all, the company's interpolation between the balé-trained dancers and the ginga-centered choreographies contributes to destabilizing the fixed dichotomy historically allocated to emblematic figurations (e.g. sexy dancing *mulatas* versus ballerinas) inside a hegemonic framework such as concert dance.

Concluding remarks

Grupo Corpo's combined use of the balé and other ethno-cultural mixtures has uniquely challenged the incompatibility that is imagined to exist between high/erudite and low/popular forms in and of Brazil. The technical mastery with which the dancers at Grupo Corpo sustain within their own bodies a frictional tension between different ways of moving and articulating ideas corporeally, as well as the effortlessness with which they transition between them, further questions the separateness that is imagined to exist between these different technologies. Above all, their performances contribute to problematizing the systematic Europeanization of Brazil's notion of citizenship (academic/erudite knowledge) against the informal Africanization of its everyday civil actions (popular/street knowledge). Swinging between movement vocabulary that have historically connoted both pride and shame, Grupo Corpo's ephemeral bricolages – especially the fixed ideas that they break down and reassemble into multiple efforts – problematize the ontological battle over which form of aesthetic knowledge Brazilians should exclusively subscribe to in their processes of self/othering identification. In addition, each time Grupo Corpo's dancers interpret Pederneiras' choreographies, the explicit pleasure associated with their hip-triggered actions as well as the memory inscribed within their muscles perform an act of revelry-revolt against the repressive coloniality associated with the ballet technique. The often seamless interpolation of get-down swings and tapping feet with upward turns and high leg lifts characteristic of Pederneiras' choreographies push dancers to overcome the hierarchy that is imagined to exist between dissimilar matrixes or ways of thinking corporeally, without forcing them to choose between the two.

It is hard to predict where Grupo Corpo is heading next in its dialogical interactions with both Brazilian artists (dancers, musicians, and set designers) and Brazil's cultural diversity at home and on global stages. What is clear is that hip-syncopation has become an essential and determining element within the company's overall aesthetic. It is also reasonable to predict that Pederneiras' lexicon of movements centered on the ginga aesthetic will continue to expand beyond the specificity it has within local/regional manifestations in Brazil, perhaps finding new sets of meaning in relation to other practices in Latin America and/or across the black Atlantic world. Since *Onqotô* (2005), for instance, there has been a subtle yet visible shift towards choreographic *apartness*, an aesthetic principle common to many dance styles in the African diaspora. On multiple occasions, dancers are given a phase or motif and are later asked to make it "their own." In this sense, it is also important to remember that each new member who enters the company undoubtedly brings to the table a new set of embodied ideas with which to articulate Pederneiras' choreographies.

Given that Pederneiras' creative process remains closely connected to the musical score, such acrobatic re-articulations will continue to reflect, first and foremost, the line of musical composers with which

Grupo Corpo chooses to collaborate. For instance, in one of the company's recent works, *Sem Mim* (*Without Me*, 2011), whose musical score is a collaboration between José Miguel Wisnik and Carlos Núñez based on the music of the Galician medieval troubadour Martín Codax, the choreography poetically reverberates a rhythmic ebb and flow of ideas between the two continents across time. Above all, the supple trajectories of the dancing bodies on stage evoke and insinuate the ephemeral yet continuous movement of the tides, the act of sailing between worlds, and the ultimate longing for adventure and to traverse back and forth across distinct realms. It thus echoes the Portuguese poet Fernando Pessoa's famous verse "navegar é preciso" ("sailing is necessary"). In a nutshell, *sem mim* stages the desire to remain in the ocean, at high seas (*mare liberum*), and to construct one's sense of selfhood in the ultimate *interzone*, or *place-in-between*. At the same time, it is also probable that Grupo Corpo's juxtaposition of exquisite ballet technique and dazzling hip swings, coated with the luminescent energy produced by restless dancing bodies, will continue to be fixed on global stages as an exoticized choreography of national identification.

All things considered, one could argue that what Grupo Corpo has been articulating in the last few decades amounts to choreographies of identification centered on an embodied mechanism that, while juggling unique and distinct concepts, seek to signify yet evade static meaning. Like other contemporary artists born or working in Brazil, Rodrigo Pederneiras has employed this particular way of organizing ideas corporeally, or *jogo de cintura*, to question and rearrange both local and foreign materials, dismantling some ties and strengthening others in a rhizome-like fashion, while improvising circumstantial bridges, juxtaposing unpredictable alliances and fragmentations, coordinating uncanny strains of thought, articulating dissonant utterances, or finding a *jeito* of composing artistic polysemy infused with serious play. A mixture of swan and she-goat. Yet, is this what it means to be Brazilian?

7
What is it about Grupo Corpo?

Case study 1: *Nazareth* (1993)

I was 18 years old when I first watched Grupo Corpo's *Nazareth* (1993). The evening began with the critically acclaimed *21* (1992) and, after a 15-minute intermission, the company premiered its latest work. Each piece consisted of almost 45 minutes of uninterrupted dancing. Though the company's innovative way of moving on stage touched me deeply, leaving a long-lasting impression in my young mind, I was equally moved to see Shell's logo splashed all over the walls of Brasília's Teatro Nacional and its lustrous program notes and posters. Certainly, Shell's substantial and unprecedented support contributed to elevating the cultural capital and the socio-economic value of a dance group whose aesthetic influences deviated from the established norm. To me at the time, the financial endorsement of the Brazilian division of an American oil company worked as an unquestionable seal of approval (see the Conclusion). As I walked out of the theater that night, I also remember thinking to myself, and later sharing with my dance teachers and fellow students, that I had never seen a professional dance company in Brazil moving "like that." Shortly after that performance, the dance critic Helena Katz proposed that: "What Grupo Corpo has been shaping, in a way that is both rigorous and singular, is a manner of dancing 'as one of them,' but using bodies as found over here" (1995, p. 23).

At that early moment in my life, watching those bodies wiggling their hips and making references to local rhythmic patterns of movement, in between precise pirouettes and sharp grand jetes, felt more like witnessing a naughty provocation against a traditional set of core values that I, a young aspiring dancer, had begun to question. Both choreographies performed that evening (*21* and *Nazareth*) made me question, intuitively, the incompatibility imagined to exist between ballet's upward linearity, geometric accuracy, and weightlessness, and other local, syncopated gestures and steps. In the midst of pleasurable sequences of civilized roguery and stylized sensuality, Grupo Corpo's performance that evening presented to me, and perhaps to an entire generation of local dancers, choreographers, and dance enthusiasts, a gateway to imagining alternative ways to conceptualize, produce, and disseminate theatrical dancing in and across Brazil.

Based on the work of Ernesto Nazareth, *Nazareth* sheds a nostalgic light over Rio de Janeiro's Belle Époque (1889–1922), when the country transitioned from an agricultural monarchy fueled by enslaved Africans to a proto-capitalist republic built on a new wave of European immigrants. Following what is now considered one of Rodrigo Pederneiras' signature elements, the choreographic score of *Nazareth* discards theatrical plots and linear narratives, and also avoids stereotypical restaging of regional dance forms. Instead, it replicates quite clearly the push-and-pull quandary written in its musical score, in this case José Miguel Wisnik's original music based on Ernesto Nazareth's Brazilian Afro-European compositions. In other words, the dance reverberates the leitmotifs as well as the formal elements, structure, arrangements, and strategies embedded within that historical musical style. In order to better grasp the ideas vivified in Pederneiras' choreographic score, below I will offer a brief discussion of Nazareth's work and Wisnik's creative process based on his musical score. Though composed for piano – a "noble" instrument (and a piece of furniture)[1] symbolically associated with European erudition–Nazareth's inebriating compositions were regarded as "popular" rather than "erudite" music during his lifetime. When compared to opera composers such as Carlos Gomes, the author of *Il Guarany* (see Chapter 6), Ernesto Nazareth was dismissed or downplayed as a composer of "black music." As Wisnik explains in the program of the ballet *Nazareth* – and further demonstrates in the DVD of its recording (2005) – Nazareth's musical legacy consists of salon music "contaminated" by local Afro-Brazilian rhythmic patterns. Although he employed the standard Western notation, Nazareth's music juxtaposes allegro and staccato music cells imported from European rhythms in fashion at the time, namely polka, schottische, mazurka, waltz, and sonata, with local syncopated cells extracted from Afro-Brazilian rhythms such as batuque, jongo, and lundu (see Chapter 3). Yet, while the hybridized melodies he composed are impregnated with Africanist aesthetic principles, the language under which he arranged the musical phrases and wrote them down is clearly European. The result is a series of unstable – or rather irreverent and irresistible – staccato-syncopated marches, counterpointed-melodies (*choros*), and maxixed-polkas, often commercialized as "Brazilian tangos." Still, despite their ascribed "inferiority," the financial success of Nazareth's music sheets meant that they were consumed widely both within working-class environments, such as popular ballroom and cinema halls, as well as European-aspiring elitist spaces of collective sociability, such as middle-class house parties and high-society balls. His ragtime compositions would, later on, have a decisive influence in the renewed urban samba culture that emerged in Rio de Janeiro during the twentieth century.

In composing the score for the ballet, Wisnik departed from Ernesto Nazareth's original scores and deconstructed them into their multiple elements. For instance, the composition employed in the first scene, "The Polka," contains the three key rhythmic cells found in "Cross, Peril!," one of Nazareth's early compositions from 1879. Briefly, the left hand sustains a European ballroom accompaniment (polka), while the right hand features a ragtime riff (maxixe) punctuated by octaved notes, in staccato (resembling Tchaikovsky's "Nutcracker"). "As in a Bachian prelude," Wisnik further explains, the composition "Cross, Peril!" "has a texture in a threefold plan, wherein one can behold at the same time the European ballroom dance, the African beat and the virtual punctuation of an orchestral nutcracker" (1993, English program). Ultimately, Wisnik's original score invites the audience to stage, inside their heads, a range of lively daring feats that swing between popular/black and erudite/white forms and contents, thus blurring these socio-culturally constructed boundaries.[2]

Wisnik's creative process is shaped up by a creative dialogue that this music composer and literature scholar formulates between Nazareth's hybrid compositions and the literary work of Machado de Assis. As Wisnik (2004) points out, like Nazareth's music, the friction between white/erudite and black/popular aesthetics in Brazil is a re-occurring theme in Machado de Assis' writings. For instance, the already mentioned short story *Terpsícore,* especially the dancer's uncanny way of dancing (1886; see Chapter 6) functions as a point of reference from which Wisnik strings together his compositions about Nazareth's music. In particular, he adopts a series of rhetorical strategies found in Machado de Assis, which he groups under the concept of "mirroring" (e.g. doubling, inverting, and multiplying) to reflect, fragment, and recompose Nazareth's original creations and its underlying themes. For instance, in the fifth scene, entitled "Baroque Sonata," Wisnik executes "Sagaz," one of Nazareth's ballroom compositions, in reverse. The result, Wisnik indicates, is a "Bachian profile" (*Nazareth* program).

Guided by Wisnik's "mirroring" deconstructions and reconstructions of Nazareth's soundscape, Pederneiras ventures, subsequently, into a direct dialogue with mechanisms available in both Nazareth's and Machado de Assis' works with which to blur the line between the white/European/erudite and the black/African/popular dichotomy. Across the ballet, more importantly, hip-swing (i.e. ginga) becomes an apparatus of mediation – or juggling mechanism – across three distinct kinds of movement vocabulary, which I hereafter identify as bird-like, samba-like, and doll-like efforts. As will be further explained below, this movement triad off-centers the dichotomy imagined between what Wisnik calls "the erudite" (i.e. European-aspiring/elitist/white) and "the popular" (i.e. local/street-like/black or mixed). In doing so, the choreography demonstrates the indefinite ways in which, in Brazilian everyday life, the "popular" is almost already "erudite" (and vice versa). Yet, as will be further discussed below, it also reaffirms how movement acquires discrete meanings according to a dancer's gender, race, sexuality, class, place of birth, etc.

In the week leading up to its premiere at Rio de Janeiro's Municipal Theater, Rodrigo Pederneiras declared to *O Globo*, one of Rio de Janeiro's major newspapers, that his contact with the music of Nazareth, especially his maxixe-like tunes, enabled him to discover the pleasure of ginga and hip-swing (*rebolado*). He also confessed that, before this montage, he was unaware that he was able to move his own hips in such a playful way (Joory, 1993). Pederneiras' personal "discovery," i.e. the sensual playfulness of his choreography, colors the uncanny arrangements structuring the musical score with light staccato steps (e.g. jolly hops), non-linear pathways (e.g. off-centering *pas-de-bourrées* and bent-elbow *port de bras*), and pantomimes of syncopation (e.g. jiggling hands). Furthermore, the use of horizontal wave patterns in the construction of the dance space, the syncopated coordination between the hands, eyes, and hips, the dancers' pleasurable attitude towards both the overall choreography and each other (e.g. smiling gaze) all dazzle the audience, producing an eye-candy effect.

While the music and dance are sassy and ebullient, the visual elements supporting *Nazareth* frame it under a black-and-white nostalgia. Zechmeister's costumes dress up dancers according a gender/class dichotomy. Men wear formal turn-of-the-twentieth-century light gray tailcoats, with matching vests, tights, suspenders (no shirt or tie), and black and white shoes. Women wear white spaghetti-strap unitards with black accents, including solid or polka dot breast cups, shoes, and turbans (or "baiana" headbands), whose ends are eventually pulled up like a unicorn horn (in the sixth scene). At times, their "white body" is adorned with black and white corsets or white see-through tutu skirts wired like hula-hoops over black underwear. As I see it, while the men's attire characterizes them as "real" elitist guest, musicians, or conductors at operas and formal balls, the women are dressed as "ideal" ballerinas or "ballerina dolls." In doing so, their costumes install a visual dichotomy between, for example, theatre-attending elites and stage dancers or music composers and their (imagined) dancing muses. Along these lines, the lighting design guides the audience to swing back and forth between two distinct scenarios across ten scenes:[3] one introspective "dream-like" or "imaginary" penumbra and another extroverted and bright "reality." Finally, the minimalist stage design, conceptualized by Paulo Pederneiras, consists of 14 three-dimensional extra-large gray roses (about five feet wide) made from pliable sheets of metal "fishnets," which are arranged into a rectangular grid in the background. Sometimes they are visible and sometimes they are in blackout.

The choreographic style of *Nazareth* draws on three major concepts that I identify here as "linear waves," "multiple voices," and the previously mentioned "mirroring." Following the first concept, for instance, the choreography moves predominantly in horizontal lines across the performative space in wave patterns. Beyond lateral entrances and exits, dancers come together in groups and dissipate in solos or duets according to this uni-dimensional oscillation. The pendular sway of dancers towards both sides gives their linear trajectories a sense of reversibility rather than a progression towards one particular direction. Diagonal and spiral pathways are employed sparsely, often indicating a dramatic disruption of the overarching flow (e.g. the entrance of Rui Moreira in the sixth scene; see discussion below). Second, the choreography vivifies distinct rhythmic cells available in each musical segment, thus reproducing multiple voices, either in solo or group formations. During the second scene, for example, male dancers are often choreographed to the piano's left-hand accompaniment (polka), while female dancers' choreography is set to the piano's right-hand melodies (classical and maxixe cells). In doing so, dancers are disconnected rhythmically according to their gender, although they share the same musical score and physical space. As will be further discussed below, different parts of their bodies may also accentuate or counterpoint one of the multiple "voices" suggested in the musical score. Hence, the choreographic organization of the dancers' body parts also reflects the concept of multiple voices.

Third, the choreography replicates the concept of "mirroring" found both in Machado de Assis' writing and Nazareth's composition in different ways. In "Flora's Sonata," for example, two female dancers mirror each other's movement in unison. In "The Maxixe," a similar mirroring effect occurs, this time between heterosexual couples, where females are also dressed in male clothes. In "Funeral March/Flora's Sonata," groups of dancers alternate between the same sequences of movement in unison. Later on, dancers move from left to right, performing the same series of movements at different times and with different intensities. In "Baroque Sonata," dancers execute the same sequence of movements at different times. Finally, in "Waltz of Mirrors," two female dancers invert each other's sequences of movements. In the ballet, this series of kaleidoscopic variations creates deep similarities or closeness between opposing surfaces, as in the two sides of a coin. It also constructs alternations between attraction and repulsion of dichotomies. In that way, the exclusionary division arbitrarily constructed between contrasting ideas (e.g. popular and erudite, men and women, black and white) ironically vanishes, as dancers appearing to be different find themselves heading towards the same direction.

Across the ballet, dancers sustain vertical postural lines characteristic of balé-trained bodies. This upward linearity is visible especially in the vertical alignment of the feet, legs, torso, neck, and head. Movements are typically instantiated from the limbs, with the torso sustaining an upright stance and the pelvis in a turn-out or "locked-in" position (*encaixado*). The legs and feet support and dislocate the weight of the body in space with balé steps modified with the suppleness characteristic of (Euro-American) modern dance. As they move, dancers lightly drag in time or bounce in space. Their gaze leans downwards to indicate introspection, forward to depict impartiality, and – more often than not – towards one another to acknowledge complicity.

The primary deviation to the dancers' "balleticized" frame or system of organization emerges at the vocabulary level: a subtle yet isolated movement of the hips between "aligned" (*encaixado*) and "misaligned" (*desencaixado*) stances. Yet, there is no attempt to permanently place the tailbone inside or outside the pelvic "case," as these expressions suggest in Portuguese. The dancer moves in and out of the "turn-out" position, ephemerally "arching" or "breaking" ballet's linearity with sideways or circular misalignments of the torso. In doing so, while the muscles around the hip joint command spiral or curvilinear rotations in space, through contract and release, the tailbone wobbles back and forth between ballet's turn-out position and a range of sideways and multi-meter misalignments. Hip-swings occur in isolation or in dialogue with other bodily parts. The subtle and sparse uses of hip-triggered swayed actions, often indicative of transgressive sensuality, are further amplified in the case of female dancers by their wobbling see-through skirts. Above all, these offbeat yet soft-spoken twists and turns choreographed from or towards the pelvic area keep the tension between the "encaixado" and "desencaixado" stances alive.

Other far more visible modifications occur at the periphery, especially within the upper body, between the round or sharp-angled upper limbs and the popping or vibrating chest. Unlike the lower limbs, arms express ideas, especially through a hand-eye coordination accentuated by facial expressions (e.g. smiles). The upper limbs, which often ornament the torso and head with art nouveau "scribbles," play a part in disrupting the linearity of neo-classical ballet or break its overarching geometry and flow. Arms (and hands) also instantiate motion in space, either replicating rhythmic cells (e.g. syncopated rhythms) or further guiding the dialogues between bodily parts. On occasion, moving arms gesticulate recognizable figures and symbols (e.g. wings). Finally, the hyperbolic rotation of the arms often compensates for the limit range of motion of the hips.

In the particular case of *Nazareth*, Pederneiras choreographs multi-meter dialogues between bodily parts, especially hand-chest, hand-feet, hand-hip-feet, and hand-eye-hips. More often than not, these bilateral and trilateral flickerings between the center and periphery push dancers to move in a polycentric and polyrhythmic fashion. Yet, although the hips are constantly coming in and out of alignment with the upper torso, the arms – and especially the right hand – dominate the articulation of ideas, whether syncopated or not. Examples include the undulated or sharp-angled fragmentation of the core as well as the round or offbeat gesticulation of limbs, especially on the right side.

In fact, the dialogue between the right bouncing fist and tapping feet (amplified by skirts), a movement that I will describe below as "jiggling hand," stands out as the primary kind of bodily syncopation, or motif, binding the choreography. The right hand pantomimes, in this case, the ways in which musicians and conductors use their predominant/rational side – usually the hands and feet – to mark or follow (complex) rhythms. Subsequently, the rationalized embodiment of syncopation tames or counterbalances the voluptuousness of hip-triggered (re) actions. At the same time, by pointing the right fist forward and wiggling it at the waistline, Pederneiras coats subliminal libidinal textures over the dancers' mimetic gestures. The superficial contact of the jiggling phallic-fist with the wobbly see-through skirt eloquently adds, one could argue, a veiled eroticism onto their suggestive hip-triggered endeavors. In sum, quivering fists and undulating skirts improvise a tantalizing construction of desire, whose rippling effects "contaminate" the entire ballet.

This over-emphasized use of the right hand to represent multi-meter arrangements in time is followed closely by the predominant use of the right foot in the oscillatory construction of space. Dancers often move their right foot back and forth, referencing the first half (or one side) of a samba step (TA, ta). In doing so, the hip joint bounces vertically and horizontally in one direction, causing the tailbone to draw a continuous figure of eight in space. As a matter of fact, the back-and-forth dialogue between the hips and feet tends to lean towards the right side almost exclusively. And, because their feet seldom shuffle back and forth as samba dancers do (see Chapter 3), this choreographic choice inhibits the hips' ability to fully syncopate. Thus, it ends up looking like a "white man's little samba" (*sambinha de branco*).

In terms of movement vocabulary, Pederneiras responds to Wisnik's deconstructions and recombinations of European ballroom beats (polka), ragtime riffs (maxixe), and staccato melodies (classical music "*a la Nutcracker*," as Wisnik puts it) with a "game of mirrors" across three discrete kinds of efforts: "bird-like," "samba-like," and "doll-like" efforts. In the first scene, each of these three efforts (which are further described below) is gradually introduced in isolation. Although they are not always synchronized with a particular musical motif, such as "polka," "maxixe," or "classical music," the choreography progressively pushes and pulls these kinetic variations within the dancers' bodies across the 40 minutes of uninterrupted choreography. By the last scenes, dancers, especially the female characters, articulate a wide range of polysemic interaction across these three kinds of efforts, embodying and disembodying what look like passive or "docilized" phallic sticks, winged figures, and hip-wobbling maxixe dancers. In a much more subtle fashion, male characters oscillate primarily between bird-like and samba-like efforts, at times interacting with female dancers and at times "dreaming" about them.

Leaning towards classical ballet, doll-like efforts are performed (almost exclusively) by female dancers. Dancers move in an angular fashion, with sharp and repeated precision, and a limited range of motion. These actions either mimic the mechanical articulation of toys, replicate the inanimate (i.e. amorphous or stiff) use of the body associated with dolls, puppets, or stick-figures, or reflect abstract ideas associated with, but not limited to, miniature ballerina dolls in musical jewelry boxes and, subsequently, maidenhood, preciousness/Europeanness, and the objectification of desire. At one point, a male dancer stands close behind a female dancer, holding her waist as she raises her phallic legs forward 90 degrees. As I see it, the "linear" quality of these actions offers citations and parodies to historical dancing scores, such as "dancing dolls" featured within European ballets (e.g. *The Nutcracker* or *Petrushka*).

Unlike the example given above, "bird-like" efforts are characterized by the soft and undulating or wing-like articulation of the arms and the playful flapping or spiral gesticulation of the hands. In a similar fashion to the "doll-like" effort, they privilege the upward and linear mobility of the core (the torso and hips), with the gravitational center of the body projected towards the chest. At times, the dancers' gestures resemble the attributes of birds (e.g. the ability to fly/glide), replicate elements within birds' quality of movements (e.g. weightlessness or evanescence) and reflect abstract ideas that may be symbolized by, but not limited to, feathered or supernatural creatures, or ethereal beings. As I see it, these efforts bring to mind historical choreographic scores, such as Loie Fuller's *Serpentine Dance* and Anna Pavlova's *The Dying Swan*.

In addition, dancers also articulate "samba-like" efforts, often accompanied by pleasurable smiles. At times, for example, they mimic or replicate the syncopated rhythms available in the music (e.g. jiggling of the right fist forward with an accompanied by wobbling of the head), resemble samba and maxixe dancing, and reflect abstract ideas that may be symbolized by, but not limited to, syncopated rhythms, which evoke blackness, sexuality, and naughtiness. The range of motions associated with the pelvic area includes rounded or sideways rotations, back-and-forth contractions and releases, and vertical bounces. Other movements associated with this effort include isolated contractions or rotations of the chest, the pelvic area, or both, alternating circular rotations of the lower arms, and bouncing steps and jumps. This "wobbling" way of moving evokes the sampling of historical choreographic scores, from references to the exoticism articulated by Josephine Baker and Eros Volusia to popular culture citations, such as "the chicken dance" and Michael Jackson's "jiggling hand" (as seen in the music videos for "Beat it" and "Thriller").

Combined, Pederneiras' playful interactions with the musical score, as well as the costume, the set, and the lighting designs, dazzle the audience with an overwhelming emotional nostalgia. In addition, the precision with which interpreters combine technical virtuosity with pleasurable enactments on stage provokes a hypnotic and uplifting effect – an eye-candy effect. Dancers oscillate with exquisite pleasure across the elevating gracefulness of fairy-like creatures, the linearity of phallic ballerina dolls, and the "get-down" indication (not imitation) of devilishness maxixeiras (naughty dancers). Resonating with Machado de Assis' depiction of Terpsichore as a "mixture of swan and she-goat," Pederneiras' choreographic mix forces the dancers' bodies to combine and juxtapose a well-defined colonial mimicry of culture and civilization (e.g. classical ballet) with an unpredictable and multiplying savagery of the lower stratum (e.g. maxixe dancing). Interwoven by soft-spoken and impish right-hand-driven conversations, these juxtaposed interactions include the dialogue amongst: (a) mischievously undulating hips, an unspoken reference to a feminized blackness; (b) pointed-toe feet, a metonym for the phallocentric-patriarch control; and (c) naughty gazes, a choreographic testament of the power of forbidden desire and the pleasure of transgression. In this particular territory of the Grupo Corpo repertoire, the choreography represents a stylized intersection of gender, race, and sexuality located within the dancers' bodies, but safely situated in the past. Finally, the specificity with which each dancer at the Grupo Corpo interprets Pederneiras' deviating accentuations compromises the uniformity expected of classical ballet in a very productive way.

Overall, the indefiniteness constructed between dichotomies and triads informs and is further formed by how blackness – both Afro-Brazilian qualities of movement and Afro-Brazilian dancing bodies – are reconfigured into the choreographic discourse, adding mew meanings to their actions. Pederneiras' choice of staging his *transcultural* score inside an elitist space could be read, generally speaking, as a reproduction of the dilemma with which Brazilian were faced at the end of the nineteenth century. But his choice of casting a black man (Rui Moreira) and a white woman (Jaqueline Gimenes) to interpret the ballet's culminating *pas de deux* clearly disrupts the invisibilization of blackness in upper-class venues.

With quotations of some of Nazareth's most popular compositions, "Rowing" and "Odeon," the sixth scene enacts a high point at a ballroom event. Towards the end of that scene, Moreira returns to the stage alone and reinstates his imposing bird-like/samba-like choreography of (forbidden) desire. This time, however, he is chased by what looks like a skirted maxixe dancer (Gimenes), identified by her lascivious hip-swings and a black horned headband. Auspicious, yet sensual, Moreira's character engages in a tantalizing pas de deux with the horned maxixeira. Although their duet does not mimic a particular ballroom style,

Moreira leads Gimenes into sinister pathways across space, hypnotizing her (and us) with his civilized grace. Unlike Moreira, Gimenes' sequence of steps combines samba-like with doll-like efforts, further offsetting the polar opposition between the two dancers towards ambivalent routes. Like Moreira's previous partner, Gimenes jumps on his lap and they hop forward in a left-to-right diagonal towards the audience, simulating a "Brazilian tango" lap dance. In the end, Gimenes' hypnotic naughtiness evokes, beyond any doubt, Machado de Assis' mixture of swan and she-goat. Holding hands as they move in half-step percussive hops, he softly pushes his hips upwards and she responds with little bounces. They smile, confirming the consummation of their transgression.

Regarding form, while the duality is evoked by the couple's respective bird-like and doll-like efforts, the samba-like effort shared between the two and their syncopated "lap dance" produces a synchronized leitmotif. In terms of content, in this pas de deux, both Moreira's supple swing of arms and hips as well as Gimenes' naughty steps and looks seem to invert the "white/male/honor"/"black /female/shame" dichotomy. Underneath the uncompromised sensuality framing the entire ballet, Pederneiras' choreography lightly touches on the bittersweet friction across the historical shame, the transgressive desire, and the physical pleasure associated with syncopated dancing rhythms. But it hides the implicit violence intersecting gender/race/sexuality relations. Pederneiras' choreographic amalgam of unlikely ideas, his "mixture of swan and she-goat," reveals subtle yet unsettling negotiations of power relations within dancing bodies. Ultimately, it gestures towards the hidden ambivalences and the dangerous ironies that the concept of "mulatto-ness" or "greyness" assumes inside a country built on both (the guilt of) slavery and (the shame of) miscegenation (Wisnik, 2004, p. 64).[4]

Following its debut in 1993, *Nazareth* continued to tour nationally across Brazil's major cities. While most reviewers responded positively to the piece, praising Grupo Corpo as the most relevant and (financially) stable dance company in Brazil, the ballet stirred up a few conflicting and critical opinions. According to Daniel Piza (2007), for example, Pederneiras resented the local press' recognition of *Nazareth* as a dismissive "sambinha de branco."[5] In spite of that, in 1994, both *21* and *Nazareth* toured internationally across Latin America, the US, and Europe. One of the highlights of this acclaimed season was the *Sixth Biennial of Dance* in Lyon, France, under the title "Mama Africa, from Africa to Harlem."[6] At the time, the organizers of the festival advertised Grupo Corpo as "one of the most important companies in the world." According to Ponzio (1994), the sophistication with which Pederneiras' choreographies (*21* and *Nazareth*) evaded the stereotypes associated with Brazilian dance, combining the technique of classical ballet with pedestrian movements and gestures from popular manifestations in Brazil, pleasantly surprised the international critics present at the festival.

Between *Nazareth* and *Breu:* a genealogy of ginga in Grupo Corpo's repertoire

When the dancers of Grupo Corpo first interpreted Pederneiras' choreographies centered on syncopated rhythms in the early 1990s, their frictional hip-swings stirred up an act of *signifying* roguery, or serious play, on stage. Ever since, Pederneiras continued to employ ginga to "arch" or "break" the geometrically elongated posture and uplifting contracted musculature associated with (neo)classical ballet. Between *Nazareth* and *Breu*, which I will discuss next, there is a visible expansion of the ways in which balé-trained bodies are pushed to interpret choreographies centered on hip-syncopation. What makes the repertoire of Grupo Corpo relevant to discussions about the geopolitics of (dancing) knowledge is precisely its orchestrated attempt to maintain the unsettling juxtapositions (power struggles) between two (or more) conflicting ways of imagining the moving body and articulating ideas. Every morning, for instance, the dancers at Grupo Corpo take part in a ballet technique class that, among other things, nurtures the alignment of the pelvis and tailbone with the core of the body (torso) in a "locked-in" position (encaixado) and the articulation of movements instantiated from the limbs (the legs and arms) outwards. Nevertheless, in their afternoon rehearsals, Pederneiras' choreographies push them to explore the pliable and isolated movement of the hips, independent and/or out of alignment with the rest of the torso (desencaixado), while also experimenting with sinuous movements instantiated from the pelvic area towards multiple directions and pathways. This particular way of moving functions, therefore, as a hyperbaton or an anastrophe, which "contaminates" the balé-trained bodies and disrupts their outward linearity with "local" colorful and accentuated elements informed by principles such as serpentine pathways, dissonance, high-affect juxtaposition, and serious play. In the end, the constant reiteration of off-centering and offbeat patterns of movement adds a coat of attractive estrangement to their balleticized dances. By the same token, the dancers employ the ballet technique, ingrained within their muscle memory, as a filtering device that distils these Afro-Brazilian qualities of movement into an auspicious palette.

In addition to taking advantage of the balletic body as an honoring framework of support for Pederneiras' artistic creations, the routine adopted by the company has kept the antagonistic tension between the two aesthetic systems physically alive. In this dynamic process, which is the driving force behind Pederneiras' signature style, while certain movements seek to encase the hips within the core frame, others seek to unleash it from that "box" (e.g. the ginga aesthetic). These back-and-forth oscillations between the turn-out position and hip-sways are implicitly indicated in *Nazareth* and, by the time we arrive at *Breu*, they become inevitable. Subsequently, their flickering oscillations have called into question the premise that ballet's turn-out position and hip-syncopation are mutually exclusive – or, as the ballet master Carlos Blasis proposed in the nineteenth century, that one is a honoring/auspicious way of dancing while the other represents a vulgar/grotesque distortion of the body.

As previously outlined, in pieces such as *21* (1992) and *Nazareth* (1993), vocabulary instantiated from the pelvic area works as (circumstantial) citations: isolated "accents" that are inserted across the choreography to either reference a local rhythm within an otherwise classical sequence of movements or to signal an auto-exotic deviation from the norm. At the time, Grupo Corpo was summarily a "neo-classical" ballet company. Hence, while the outward mobility of the dancers' balleticized bodies polished and tamed the (desired but not-honorable) hip-triggered syncopations, the (physical) centrality of the navel area made the dialogue between the hips and feet in charge of managing the postural (mis)alignments of dancers towards or away from the linearity of classical ballet. Gradually, ginga became a vital tool or mechanism of manipulation through which Pederneiras connects music and dance. Within pieces such as *Sete ou Outo Peças Para um Balé* (1994), for example, the choreography dialogues with Philip Glass' minimalist composition, adding contrasting or complementary rhythms to the audible score, as recorded by the group Uakiti. In sum, in this phase, ginga functions as a metonymic device that either corporeally reverberates or counterpoints the interactions inscribed within the audible score.

In *Bach* (1996), whose original score by Marco Antonio Guimarães (from Uakti) dialogues with the work of J.S. Bach, the high-juxtaposed stances and concentric flow characteristic of the ginga aesthetic gesture towards the specificity of Brazil's colonial-baroque heritage.

In *Parabelo* (1997), composed by Tom Zé and José Miguel Wisnik, angular dissonances and serpentine pathways re-capture elements of Brazil's northeast regionalism. Then, in *Benguele* (1998), the tapping of feet, stomping shoulders, and the fragmented syncopation of the hips reverberate the rhythmic references to African ancestry already embedded in João Bosco's music score. Across these choreographies, ginga becomes the most identifiable feature of Grupo Corpo's signature style: a common denominator employed to articulate a multitude of (heterogeneous) embodied ideas deeply grounded in the specificity of the Brazilian society. In addition, in all these choreographies, Pederneiras employs hip-trigged actions to choreograph an overt celebration of sensuality explicitly connected to Brazil's socio-cultural context and implicitly linked to its biopolitics. Eventually, the physicality of Grupo Corpo's balé-trained bodies – and the honoring structure they provide – works as a framing device with which to valorize (and validate) the strangeness of the ginga aesthetic within the context of concert dance. Once read as a grotesque and "dishonest" way of moving, in this renewed space, the ginga aesthetic (and perhaps even hip-wiggling) is received as a source of pride across the nation.

Since 2000, Grupo Corpo has benefited from an exclusive maintenance sponsorship by Petrobrás, Brazil's state oil company, via the Rouanet Law (see the Conclusion). As a result of this new partnership, the company has begun to produce a new ballet every other year. Beginning this new phase, *O Corpo* (2000) takes advantage of Arnaldo Antunes' neo-concrete play of bite-size words/sounds to explore staccato and sharp-edged mediations around bodies, technology, and post-modernity. Here, ginga functions as a tool of mediation between the cut-and-dried movements that bounce towards and away from the ground as well as the urban scenario that the music creates.

In the twentieth-first century, the specificity of Pederneiras' situated knowledge – both as a Brazilian dancer trained in classical ballet and as the resident choreographer of a company based in Belo Horizonte (Minas Gerais) – continues to play a decisive role within Grupo Corpo's repertoire. In more recent productions, however, he has stepped beyond his particularized use of ginga as a dynamic mechanism that gestures towards the specificity of Brazilian culture. In these newer choreographies, ginga-driven sequences reflect broader (cross-cultural) themes related to intersected gender and sexuality more broadly. In *Santagustin* (2002), inspired by the work of St Augustine, ginga becomes a humorous figure of speech to address ideas about love, eroticism, and the earthly pleasures of the flesh since medieval times. Furthermore, in *Leucona* (2004), ginga ties the tangos of Cuban composer Ernesto Leucona (1895–1963) to the choreography of passion associated with Latin America ballroom dances. In *Oncoto* (2005), the sensuality of ginga-centered undulations weaves sublimated and erotic metaphors about creation, procreation, and "scenes of origins." In *Breu* (2007), which I will discuss next, Pederneiras employs ginga as a rhetorical strategy to choreograph risk-taking instabilities associated with urban/contemporary violence.

Case study 2: *Breu* (2007)

I first saw *Breu* (2007) at Grupo Corpo's headquarters in Belo Horizonte during rehearsals in August 2008.[7] At the time, the company was getting ready to take it on the road and this was the first time they would run it since their previous season. After their usual ballet class, the dancers went backstage to put on kneepads, rubber shoes, and long-sleeved shirts to protect their skin. I remember noticing an abrupt change of pace. The lively and playful atmosphere carried out since 9 am had vanished, leaving a weary sense of hesitation on their faces. They came back slothfully and gradually lay on the ground, on their marks, in a manner that could almost be read as a protest. They waited in silence, amorphous, on the ground. Sensing an eerie tension creeping up, one of the dancers cracked a joke. With a quasi-Spanish accent, he shouted with a melodramatic conviction, "hay que endurecerse, pero sin perder la ternura jamás" – Che Guevara's infamous motto "one must harden without ever losing tenderness." His remark was followed by a burst of giggles across the stage that certainly released some of the tension that had built up in the air. As I look back at that moment, I remain astounded by the accuracy with which the impromptu punchline synthesized the "subtext" rolling through the dancers' minds. *Breu* is, as I was about to find out, physically and psychologically painful to the point of exhaustion. Deep breaths. Music starts.

In *Breu*, which means darkness, the stage box is transformed into what looks like a black-tiled bathroom, where over-stimulated "zombie-like" bodies undertake abrasive dialogues instantiated by jerky hip-triggered events. Composed by Lenine,[8] the musical score sustains a circum-Atlantic bricolage of military brass bands, carnival marching trombones, *cabloquinho*[9] drums, medieval flutes, and heavy metal guitars. At its heart, Lenine's soundscape pays homage to the centenary of *frevo*, a fast-paced syncopated dancing rhythm historically connected to choreographic battles staged by capoeira players during carnival on the streets of the twin port cities of Olinda and Recife in the state of

Pernambuco.[10] Lenine's carefully constructed cacophony – a kind of Sepultura meets Dante Alighieri on Mardi Gras Day – offers a connection between urban/modern brutality and the epistemic violence of colonial and post-colonial encounters staged in Brazil. When superimposed on the lighting, the set, and the costume designs, Lenine's multi-layered citations of historical frevo compositions against electric guitars and other sharp sounds contribute to situating *Breu* in a post-modern purgatory, where repressed fears and illicit desires of a society with a dark past are let loose.

In *Breu*, Lenine's commissioned soundtrack functions as the ground zero or driving source for the ideas replicated or reflected in the choreography. However, unlike many of Grupo Corpo's earlier spectacles, here the relationship between the musical score and the choreographed script is not limited to an illustration of or a direct dialogue with the musical score. Instead, the soundtrack functions as an audible scenario, which situates the choreographic discourse in the intersection of overlapping contexts (Brazil's urban life, street parades in Pernambuco, international violence, colonial brutality, etc.). In "Secular," a climactic moment in the musical score, Lenine orchestrates a homage to the centenary of frevo in collaboration with the *Spok Frevo Orquestra*. This feverish textural motif, filled with both historical citations and colorful improvisations, reverberates across the entire musical score. Subsequently, the choreography responds with fragmented and juxtaposed motifs that insinuate frevo's movement vocabulary or its context. Since frevo has functioned as a source of inspiration for both the musical and choreographic scores, a brief overview of this dancing rhythm may help us to elucidate some of the thematic efforts articulated across the spectacle.

Meaning "boiling point," frevo is a transcultural dancing rhythm historically associated with working-class street carnival parades in the state of Pernambuco (Cassoli, Falcão and Aguiar, 2007). Executed with frenetic vigor, frevo's acrobatic feats sometimes resemble the maneuvers of Barynya male dancers (a Russian folk dance). Danced initially by men only, to the point of exhaustion, frevo mixed fast-paced and syncopated footwork with leg crossing, bouncy squats, forceful punches, acrobatic inversions, and an intricate set of jerky outward gestures accompanied by small umbrellas as balancing props.

As most ethno-musicologists and historians seem to agree, this way of acting in a mob towards a boiling point gained wider visibility during the last decades of the nineteenth century, as military brass bands enacted musical battles on the streets of Recife and Olinda during carnival festivities. Because the confrontation would often become physical, bands such as the historical *Quarto Batalhão de Artilharia*, known as "O Quarto" (The Fourth), and *Corpo da Guarda*, known as "Espanha" (Spain), featured local capoeira players who were hired to "dance" in front of their brass bands, aggressively pushing their way through the throng of revelers. Mixing aggression with playfulness, these dancing "bodyguards" engaged in improvised choreographies of violence that incorporated bats, parasols, and razor blades into "stylized" capoeira steps and gestures. After the abolition of slavery (1888), as the racial tensions rose, the number of working-class parading troupes featuring frevo's frenetic rhythms and aggressive (capoeira-like) movements in public spaces also increased. As an animated mob of black and mixed-race men forced its entrance across packed crowds or met with a competing brass band, it was not unusual for their spectacular display of virility to boil down to bloody fights. Similar to the batuque parades during colonial Brazil (J.J. Reis, 2005; see Chapter 1), these brass band dancing parades appropriated the public (festive) space to articulate their processes of cultural resistance/existence (counter-narratives). Hence, the martial aggression woven into the soundtrack and replicated in their improvised choreographies fed off and reverberated the brutal negotiations of race relations enacted in these port cities in the aftermath of the abolition of slavery. In the following decades, the police brutality associated with the persecution of these Afro-Brazilian brass band parades and other "uncivilized" practice that presented a threat to the society played a decisive role in the rapid transformation or, rather, "docilization" of the "old school" ragged and violent steps into a joyful and acrobatic dance form as frevo is known today.[11]

Departing from Lenine's citations of historical frevo melodies, as well as his appropriation of the boiling effervescence of Pernambuco's brass-band style of playing frevo music (call-and-response music battles), Pederneiras employs polycentric and polyrhythmic qualities of movement to choreograph risk-taking instabilities associated with urban violence in places like (but not limited to) Brazil. In particular, the choreography weaves together a series of identifiable qualities of movements extracted from frevo dancing, including its frenetic syncopation with rapid change of levels, transversal balances, inverted positions, push-ups, and squat work. Hence, instead of mimicking or faithfully reproducing particular units of movement associated with contemporary frevo dancing, the choreography replicates a wide range of historical ideas associated with the form, the content, and the context of frevo dancing in order to talk about the chaotic times in which we live. Broken into ebullient fragments and danced largely on the ground, these patterns of movement conjure up different aspects of asymmetrical power relations.

Throughout *Breu,* the dynamic conversations between the hips and feet no longer stress the soft-spoken sensuality and the smooth playfulness which had been identified as one of Grupo Corpo's signatures since *21* and *Nazareth*. More specifically, Pederneiras trades the tongue-and-cheek "sensuality" of lascivious ginga for the virilization of hip-triggered endeavors (characteristic of Afro-Brazilian practices such as capoeira and frevo) to evoke concepts such as aggression, violation, volatization, and exhaustion. Subsequently, the use of ginga as an off-centering dynamic – or unbalanced-balance – instantiates a series of (purgatory-like) battles of conflicting desires between passive and aggressive interactions or between free falls and collisions. In this scenario, attraction and repulsion, dependence and nihilism, or cruelty and passion are not necessarily distinguishable.

The frictional mixture of euphoria and torment portrayed in *Breu* begins with its choice of name. In addition to darkness, the word "breu" references a type of rosin with which ballerinas rub their shoes to avoid slipping. Nevertheless, in the poster created for the ballet, the title of the piece is "written" with a white powder, arranged on the top of a mirror-like surface. As I see it, the typography references *Cosmococa* (1973), a historical installation/collaboration between Hélio Oiticica, Jimi Hendrix, and Neville D'Almeida. In this photo installation, the artists enhance the facial features of iconic celebrities (e.g. Richard Nixon and Marilyn Monroe) with lines made with of what looks like cocaine powder. Unlike the ballerinas' rosin, a frictional powder that prevents dancers from losing their poise by either holding their posture or keeping them from sliding as they move, cocaine powder is an anesthetic-stimulant dancing drug. Among other things, it leads moving bodies to remain in an artificial state of agitation or even to move with an ebullient frenzy.

As the piece begins, Paulo Pederneiras' lighting and set designs transform the performative space into a large room furnished with (what looks like) black ceramic tiles. Dark and reflective, it consists of a grid of about 1,800 black fiberglass square panels (40 x 40 cm) that tiles the back and sidewalls with a semi-glossy and refractive texture that absorbs and partially refracts glimpses of the dancers as they move. A black and glossy linoleum floor covers the stage that, when lit, creates the illusion of an inverted underground below the dancers' feet. Far from representing the vertical bottom of the stage, the linoleum's mirror-like surface and its reflected lights transform the stage floor into a vertical portal between two realms connected vertically by a symmetric inversion. As I see it, the coldness of this bathroom-like environment alludes to a public (doomed) territory, a shelter permissive of illicit or violating endeavors.[12] Contrary to expectations, the minimalism informing this dark public "bathroom" functions as an amplifier of darkness.

Freusa Zechmeister's costume design is equally minimalist. All dancers wear shiny and close-fitting unitards, whose black-and-white fabric patterns vary from solid colors to zebra-like stripes or electrical wiring grid patterns. Like the costume of a carnival jester, the juxtaposition of different fabric patterns in a single unitard visually fragments the dancers' bodies into distinct sections, such as left and right sides or core and limbs. Likewise, most dancers' unitards feature a solid black back, further dividing the moving body into contrasting "animated" front and "somber" back halves. Every time they turn to the back wall, more importantly, their costumes reduce their body mass to floating silhouettes in the dark, thus temporarily erasing their tangible physicality from the stage. At the same time, the lighting design deepens their volatile presence, imprinting ghost-like reflections over the tiled background as well as inversely extending their existence underground below the reflexive linoleum floor. At last, their make-up consists of a one-inch-thick white stripe across their eyebrows, contrasted by black lipstick and eyeshadow. While the dark eyes and mouth add a grave tone to their expressions, the white stripe artificially illuminates their foreheads, thus masking any physical trace of tension or concern. The result is a mob of cold-blooded impartiality.

Like *Nazareth, Breu* makes full use of *linear wave patterns (pendulum)*, where the choreography groups and dissipates dancers horizontally in space; *multiple voices* (*fugue*), where the choreography arranges different (groups of) dancers in space and time, according to multiple ideas available in the musical score; and *mirroring patterns*, where the choreography duplicates, inverts, or multiplies one unit of movement across the stage, varying its intensity, frequency, direction, etc. Aside from these reoccurring patterns, the two major innovations within Pederneiras' use of the choreographic space are the increased use of diagonals, especially from right to left, and the valorization of the horizontal space situated directly above the ground.

The choreographic score reflects or counterpoints a range of elements available in Lenine's quilted soundscapes (e.g. riffs of historical carnival marches, urban/mechanical sounds, and "samples" of heavy breathing). And, though the audio and kinetic ideas are not always synchronized, the choreographer translates these audible ideas into kinesthetic metaphors (e.g. risk and vertigo). In "Trovador" (the sixth scene), for example, a couple (Flavia Couret and João Vicente in the original cast) enacts an abrasive entanglement on the ground (upstage right) with abrupt and explosive interactions, whose drive and cadence are rhythmically dissonant with the troubadour's soothing medieval flute percolating the scene. Their duet is also kinesthetically distinct from the dancers, who glide around the stage in pairs, like brothers-in-arm carrying corpses from a battlefield. In this visceral "love-fight" pas de deux, a highlight of the ballet, the duet snaps out of the musical cue with an erotic violence that drastically contrasts with the scene's dismal melody and its choreographed melancholia.

On the vocabulary level, Pederneiras' interpretation of the musical score includes (but is not limited to) a reflection about the choreographed ideas articulated by: (a) groups of singularized dancing bodies who share the performative space, while maintaining their personal style of dancing apart; (b) restless bodies, whose exhaustion becomes, at the same time, a goal and a driving force; (c) risk-taking bodies, leaping in space towards the ground or pivoting sideways fearlessly; (d) volatile bodies, jerking between attraction and repulsion, revolt and revelry; (e) marginalized bodies, whose mobility and function make them disposable or unaccountable matter; (f) bodies-at-risk, almost lifeless, dragged with compassion (or guilt) by their heavy limbs, shoulders, ankles, chins, or hips.

Regarding the use of the body, in *Breu* nearly all innovations may be traced back to one strategy: sending dancers into horizontal positions on and off the ground.[13] Turning his investigation of hip-driven movements sideways, Pederneiras' productive battle with gravity choreographs daring feats of ephemeral beauty, instantly cut with aggressive bursts and morbid slams of corpse-like falling matter. Choreographed mostly on or near the floor, dancers push their hips upward, seeking desperately to rise horizontally, but inevitably sinking back. The overall choreography results in a desperate sense of vertigo, punctuated by an "aesthetic of brutality" reflected, for instance, by fork-like leaps that intersect dancers' mid-sections, dancing torsos that slam against one another, followed by jerky embraces and, as Katz (2007a) suggests, phallic legs piercing upwards in the air.

Danced parallel or attached to the ground, *Breu*'s overarching horizontality causes dancers to reconfigure the division of labor between their body parts. The contractions and releases of muscles located in the pelvic/stomach area become the centralized force responsible for the body's ability to dislocate in time and space. Moving in and out of sideways stances, the upper limbs assume the role generally expected of dancing feet: that of supporters and/or propellers of bodily weight towards or away from the ground. When the dancers' arms and hands are not directly touching the ground, either catching or propelling them in motion, they use their upper limbs to counterbalance the direction and the intensity of their dancing efforts. With these new functions, the hands and arms rarely perform ornamental gestures. By contrast, the legs and feet are constantly being tossed in the air, embracing and punching other dancers, or even articulating ideas. Dancers wave their legs like zebras, jumping off the ground in inverted positions or piercing the void upwards with determination. Following the choreographic intention to remain near the ground, the movements generated from the hips upward cause the core of the body to flip on its own axis, while transitioning to different horizontal positions. Furthermore, as dancers try to "get off the ground," their hips rise up with jerky acute motions, pushing the rest of the body horizontally in the air.

As already tested in *Oncotô* (2005), here the choreography is not interested in overcoming gravity. Instead of denying the overwhelming force that limits and defines their range of motion, for example, dancers throw themselves in the air, bouncing off the ground or free-falling towards it, while making the weight of their bodies visible and audible. Far from being ethereal, dancers enunciate "off the ground" leaps and slings with the experiential certainty that they will inevitably go back to it. In addition, these transgressive bodies articulate asymmetrical negotiations that seek to push movement boundaries, but often also risk hitting their kneepads on the linoleum floor. In the case of the "phallic leg" movement (see Figure 7.1 below), for instance, the hips propel the pointed feet and the straightened legs upward and forward like a wave, then the feet land on the ground with an audible slap. The noise is so loud that these repetitive actions become percussive.

The choreography at times reflects efforts associated with the violent negotiations of power relations of urban modern life and, at other times, replicates efforts associated with particular movement practices embedded into the musical score (e.g. heavy metal). On a number of occasions, the dancers execute virile actions with a certain raw urgency and determinism that ultimately redirect the aggression and/or violation towards their own bodies. This strategy seems in tune with groundbreaking movement research carried out by other contemporary dance companies from Brazil, such as Grupo Cena 11.[14] However, unlike Grupo Cena 11, Grupo Corpo's choreography seems to be invested in the visual effects of daring movements rather than its underlying perception and mechanisms. Dancers grind the hip joints towards the ground abruptly, shifting between high-affect juxtapositions that accentuate angular fragmentations of the core and scraping lacerations of its parts. Instead of swaying from side to side with a flirtatious attitude, dancers pound their joints like a pestle, crushing seeds in a mortar with repetitive willpower. In the end, the audience is assaulted by the violence with which the dancers violate each other's kinesphere, using their own weight as a weapon. These movements amount to a different type of friction. I propose that this aggressive syncopation reflects or reinstates the repeated and unwanted bodily violation characteristic of torture, rape, or self-flagellation.

Exhaustion is another key idea that leads dancers to push themselves to their physical limits on stage. Physically woven into the choreography as a rhetorical device, Pederneiras' boiling discourse drives dancers towards the compulsive weariness of their bodily resources. Between sets, recorded and live heavy breathing references actions that seem out of place. Nevertheless, upon a closer look, it becomes clear that this fervent expenditure of physical resources in *Breu* also gestures towards the improvised choreographies performed across the circum-Atlantic world during carnival and other popular festivities. Far from being pure nonsense, exhaustion should be viewed here as a reoccurring theme indicative of the historical appropriation of the public (festive) space by marginalized groups in port cities such as Salvador, Recife, New Orleans, and Port de Prince and its performative acts of existence/resistance. Whether viewed as an evil threat or as a harmless escape-valve, in these revelry-revolt acts, exhaustion fulfills a variety of functions, from physical gratification to metaphysical offering, all of which end up, at some stage, transforming pleasure into unbearable pain (and vice versa). Executed without hesitation, the firm and explosive steps in *Breu* evoke, for example, New Orleans' Mardi Gras motto "feet don't fail me now." And, as scholar Joseph Roach inquired about the devotion of the New Orleans' Black Mardi Gras Indians, "for whom do you sweat?"[15]

Ultimately, the exhaustion that the dancers endure across the choreography becomes a central metaphor in *Breu*. In this glossy bathroom-purgatory, there is no dramatic begging and suffering. The physicality of the dancers, their muscles and joints, generates both the motions and emotions on stage. In particular, movements that evoke jerkiness are executed with such an athletic virtuosity and a cold-blooded impartiality that it gives the impression that they are artificially animated or destitute.

In terms of vocabulary, two thematic movements that percolate the entire ballet, which I identify as "rowing" and "crab walking" (a kind of frevo step danced on the ground; see Figure 7.1), offer good examples of how hip-syncopation stages violence. In the case of "rowing," dancers lay on the ground with their bellies up and their arms bent outwards at chest level. With repetitive circular motions, they slide across the floor by pushing their upper arms against the ground with their elbows sticking out, as if they were "rowing" from top to bottom. Once their elbows reach their waist level, they circle up their elbows until they reach the ground near their heads, to restart the pattern. As they press their triceps down against the ground, they shoot their hips and lower limbs straight up in the air in a diagonal line. Collectively, this repetitive action resembles a series of zebra-like phallic limbs (Katz, 2007a), piercing the void with pointed rubber shoes. In the "crab walking," dancers sit in a crossed-leg position, leaning their torso back and their bodily weight with straight arms along their side. With flat palms on the floor and fingers spread out, the dancers lift their buttocks off the ground and "walk" on their hands, flapping their knee-bent crossed legs from side to side. Upon a closer look, one notices that this awkward "crab walking" forward evokes a modified version of the basic frevo step (*passo*) close to the floor. In both cases described above, these movements privilege the space right above the ground and horizontality.

Finally, it is possible to also notice an increasing differentiation of interpretations between dancers in a group, which introduces accentuated variations within the final performance. Confirming this interpretation, in an interview with the *Estado de São Paulo* newspaper, Pederneiras affirmed that in *Breu*, his main concern was related to the "raw" intent of the dancers rather than with presenting a polished final product (Katz, 2007a). Primarily, his choreography valorizes the intent and function before form, especially when the dancers are within group formations. These group movements are not necessarily performed with precise synchronicity. Instead, the ability to flip one's body sideways on the ground using only the rotation of the hips (function) takes precedence over the exact position that the legs and arms must assume during that process (form).

Breu could be read at different levels.[16] Like *Nazareth*, it also employs pathos as a choreographic figure. In this case, however, the violence embedded in the choreography produces a kind of "Brechtian" V-effect, which combines beauty with brutality. As the program states, *Breu* is a story of (post-)modern urban violence. As I see it, this aggressive and ebullient choreography reveals a side of Brazil that is not exotic

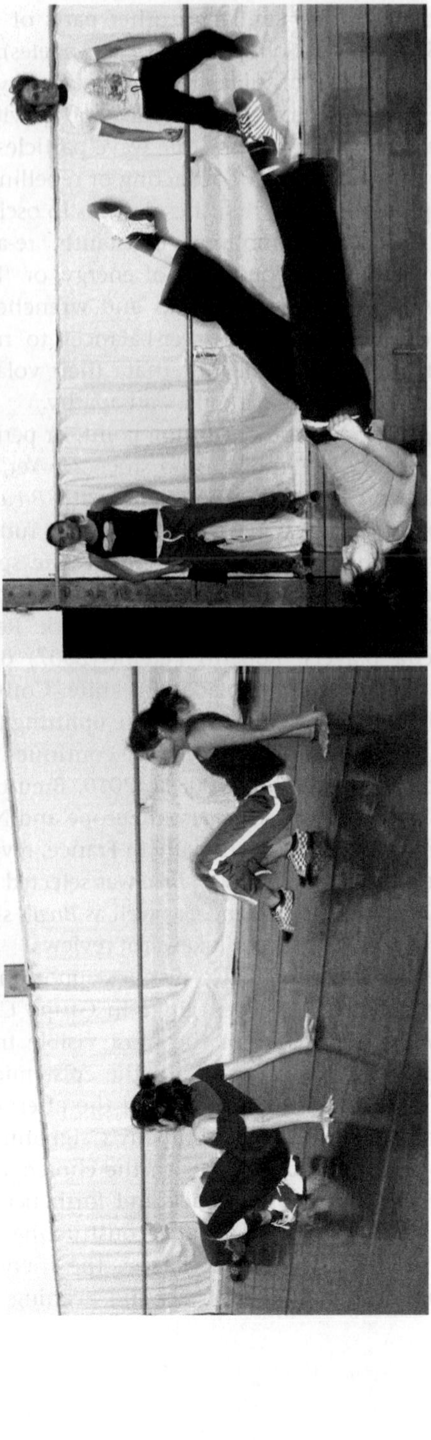

Figure 7.1 Rehearsal of Grupo Corpo's *Breu* © Cristina Rosa, 2008

or exceptional, but is rather similar to other parts of the globe (e.g. the outskirts of Shanghai or South Central Los Angeles). In that sense, *Breu* portrays volatile scenes of subjugation populated by bodies at risk. Contrary to the mathematical predictability of gravitational fields, their movements resemble quantum scale wave particles that have not decided whether they are better off attracting or repelling one another. As the choreography progresses, it pushes dancers to oscillate randomly between passive and active motions, constantly re-adjusting their precariousness. With unstable outbursts of energy, or "bipolar" intentionality, dancers flip from frenetic jerks and wrenches to complete meltdown, or from dependable counter-balances to ruthless projections. And, most importantly, they alternate their volatile instability between stiff actions, corpse-like inertia, and apathy.

Breu may be regarded as another turning point, or perhaps an exception, within the history of Grupo Corpo since *21*. Yet, it is too early to predict or reflect on how the experiments with *Breu* might further impact the company's future trajectories in the long run. Across Brazil, and especially within the Rio–São Paulo circuit, the specialized press has reviewed *Breu* with positive comments. According to Katz (2007a), *Breu* represents Grupo Corpo's most political work. Referring to the "old-time" capoeira players, Roberto Pereira (2007) defines *Breu* as cutting, dry, and precise, like the blade of a knife. Conversely, for the mainstream local audience, accustomed to the uplifting tones of Grupo Corpo's previous works, this particular dance continues to be regarded as a difficult piece to digest. From 2008 to 2010, Breu toured internationally, performing mostly in cities across Europe and North America. Within Europe and Canada, and especially in France, reviews presented *Breu* with positive enthusiasm. In 2008, *Breu* was selected to open BAM's season. Nevertheless, this performance, as well as *Breu*'s short-lived tour of the US, was met with sparse and lukewarm reviews.

In sum, *Nazareth* and *Breu* offer two distinct opportunities to understand how Pederneiras has deployed ginga in Grupo Corpo's dances. Between *Nazareth* and *Breu,* one of the most visible transformations within the company has been the size of the epistemic gap between their morning classes (ballet technique) and the afternoon rehearsals (Pederneiras' choreographies). The company's signature style results from the collaborative solutions that both the choreographer and the interpreters offer in order to jump back and forth between the two. Hence, rather than deconstructing and reconstructing one particular language (morning ballet technique classes), the frictional apartness presented on stage (afternoon rehearsals and evening performances) extend the constant back-and-forth negotiation between the two systems. Pederneiras has addressed a number of issues pertaining to the specificity of Brazilian culture in his creations. In the end, though, to say that his unique approach to these national issues and his personal choreographic style represent dancing that is "precisely Brazilian" is problematic, to say the least.

Conclusion: The Pride and Shame of Being a Swing Nation

Introduction

On Friday, August 15, 2014, I woke up at a small farm located in Brazil's Atlantic forest, about three hours south of Salvador. The *Kilombo Tenondé*[1] is a cultural and ecological center located near Bonfim, a small settlement close to the coastal town of Valença. It consists of an 80-acre farm built and sustained by the capoeira angola Mestre Cobra Mansa.[2] I will return to this farm by the end of the book, but this is what lay ahead of me.

To begin with, my time at the Kilombo coincided with the end of a three-month journey across Brazil, where I hoped to finish the manuscript of this book. From that farm, my original plan was to travel northwest to catch the end of the Sisterhood of Boa Morte's Festival (*Festa da Boa Morte*), then return to Salvador and catch a flight back to the US. As mentioned in Chapters 3 and 5, the Festa de Boa Morte is an Afro-Catholic religious celebration that takes place in Cachoeira every year, from August 13 to 16. Meaning "waterfall," Cachoeira is a picturesque city by the river filled with colonial houses painted in bright colors. It is located in the countryside of Bahia (*recôncavo baiano*) to the west of Salvador. It is also a calm and vibrating place, surrounded by hidden Candomblé temples protected by luscious vegetation. It seemed an ideal place to close this chapter of my life.

It is worth noting that Cachoeira had been a fundamental point of reference in the multi-sited field research I conducted for this book. The Festa de Boa Morte, for instance, was a perfect place to meet the elder sambadeiras and dance samba de roda, participate in open rodas of capoeira angola, visit Afro-Brazilian religious temples and attend their ceremonies, watch centuries-old hybrid processions that mix Catholicism with

Candomblé, and eat Afro-Brazilian food cooked and sold on the streets by "real baianas." It was also in that small village that I began to really understand the extent to which the ginga aesthetic had influenced the way in which people in Brazil take care of themselves, relate to others, move around the world, and make sense of it. As I strolled through its little alleys and talked to elder black women – larger-than-life figures such as the already mentioned sambadeira D. Dalva (see Chapter 3) and the late priestess D. Nilta from the Angola Candomblé temple *Nkosi Mukumbi Dendezeiro* – that I began to understand it all.

It was also there that I began to realize how Brazil's pride-and-shame conundrum was historically formed and how it managed to outlast the end of colonialism and slavery. Cachoeira was once a rich port city, with a history that was similar perhaps to New Orleans. Later on, throughout the nineteenth century, it turned into a refuge for Afro-Brazilian communities running away from Salvador, where the persecution against their religious practices had intensified. More recently, Cachoeira became a kind of celebrated "missing link" between Brazil's urban black cultural identity and its "roots" (i.e. the myth of Africa) – a place where Afro-Brazilians recuperated-cum-invented their "imagined communities" in the diaspora. But I still had many unanswered questions about that place, hence my desire to return to that site.

It is equally important to know, as I will further explain below, that Brazil has gone through a dramatic set of transformations in the last 20–30 years. The country's economy had grown and its production in the field of arts and culture had also flourished in multiple directions. Yet, here is the bottom line: despite all the changes, some of Brazil's underlying problems and paradoxes remained unresolved. Its lasting pride-and-shame conundrum was one such dilemma. Why wasn't Brazil's political emancipation and economic development followed by the settlement of its ethno-cultural quandary? What changed and what remained the same? What happened to the ginga aesthetic in all of this? And, more practically, where should I go to find the answer to these questions?

One week prior to my trip to the heart of Bahia's recôncavo and the Festa da Boa Morte, I learned that the dance department (*Escola de Dança*) at the Federal University of Bahia (UFBA) was hosting a two-day conference on dance education in Salvador during that same weekend (August 16–17). The conference, one of the other guests at the farm told me, was the "opening act" of the National Workshop of Contemporary Dance (*Oficina Nacional de Dança Contemporânea*), a week-long dance festival featuring local groups and national headlines such as the Rio-based choreographer Deborah Colker. Both events were a joint effort organized by the three major public institutions related to dance in Salvador: UFBA's dance department, the dance school of the Cultural Foundation of the State of Bahia (FUNCEB), and the Castro Alves Theater's dance company Balé Teatro Castro Alves (BTCA). Together, they offered a plethora of movement classes, lectures, round-table discussions, and both site-specific and concert dance performances in many distinct genres. Free and open to the general public, these initiatives belonged to an ongoing attempt to rescue and re-situate Salvador as a major dance destination or "City of Dance," outside the Rio–São Paulo cultural axis. In Salvador's favor, up to the 1980s, UFBA was the only university in all of Brazil to offer an accredited dance program.

When I woke up on the morning of August 15, I felt as if these equally valid opportunities were pulling me in opposite directions, hence increasing my anxiety. As was customary, I got up at dawn to train capoeira with other guests at the farm. As I began to sway my body from side to side, my mind continued to oscillate, to and fro, between these multiple propositions. What do I want: modern cityscape (Salvador) or colonial countryside (Cachoeira)? Or should I stay put at the Kilombo and just focus on the writing? If my project were a dance, how should I end it? After much sweating and shifting of intentions, I decided to go back to attend UFBA's dance conference in Salvador.

When I got there, I learned about two other cultural events that attracted my curiosity. On the one hand, there was the *Agosto da Capoeira,* a month-long capoeira festival held at the Fortress of Capoeira.[3] Similar to the dance conference/festival I hoped to attend, all activities in this mega event were free and open to the general public. On August 16 alone, the Fortress offered a total of six movement workshops of capoeira (mostly regional style), three master classes of Afro-Brazilian folkloric dance (*dança afro*), two percussion workshops, and one lecture, followed by a samba-reggae concert at night. Parallel to these events, at around 1 pm that same day, religious figures and devotees of Candomblé were scheduled to march from Salvador's historical downtown to the Church of St Lazarus and St Roch, on top of the Federação neighborhood.[4]

On August 16, I got up and, after much hesitation, headed to the dance educators' conference. It began with a Laban-based studio workshop at UFBA's campus in Ondina and ended with a series of site-specific performances in Salvador's historical downtown. To be clear, this multitude of cultural offerings was not unique to Salvador, nor was it particular of the month of August. As I said above, things had really changed in Brazil in the last few decades. As I walked to the university, a pressing question popped into my mind: how did we (Brazilians) get here and, again, where is ginga in all this? My answer to this complex question is informed by three key sources: (1) *Nova Lei Rouanet* (2010), a report produced by the Ministry of Culture and published online[5] (hereinafter referred to as 'MinC, 2010'); (2) *Lei Rouanet Percursos e Relatos* (2011),[6] a dossier organized by A.C. Abdalla, containing a collection of essays and newspaper articles on the development of the Lei Rouanet (hereinafter referred to as 'Abdalla, 2011'); (3) a conversation with Helena Katz, professor at PUC-SP (Skype interview in August 2013, hereinafter referred to as 'Katz, 2013'). Together they provide valuable data to my conclusions.

As I will unpack below, following the re-democratization of the country in 1986, Brazil went through a complex process of transformation led largely by shifts in its public policies toward arts and culture and a socio-economic boom. Amongst these factors, it is worth noting chronologically: (a) the separation of the Ministry of Culture from the Ministry of Education in 1986 and the creation of new subordinate institutions designed to promote and preserve, for instance, Afro-Brazilian cultural heritage (e.g. *Fundação Palmares*, created in 1988); (b) the gradual expansion of institutions of higher education since 1986, with programs related to the education, production, and management of the arts, including dance; (c) the implementation of a series of incentive laws geared towards cultural projects at the national, state, and municipal levels (e.g. the Cultural Incentive Law, instituted in 1991); (d) the institutionalization of a program, following UNESCO's guidelines, to safeguard Brazil's intangible heritage (*Programa Nacional do Patrimônio Imaterial*, instituted in 2000), including distinct kinds of knowledges, forms of expressions, and ceremonies addressed in this book; and finally (e) the implementation of a new piece of legislation, which makes the teaching of African and Afro-Brazilian history and culture in primary and secondary schools compulsory (Law 10.639, instituted in 2003).

Background: rebuilding a nation

Brazil's recent military dictatorship (1964–85) was a period punctuated by the imprisonment, torture, and/or deportation of artists and intellectuals who opposed the regime. The military regime also crushed all embryotic attempts to expand and diversify the country's educational system beyond mainstream/Eurocentric models.[7] At the same time, while central government repressed counter-culture movements and initiatives, it invested in the development of a unifying infrastructure of telecommunication managed by state-owned companies, e.g. *Telebrás* and *Embratel* (Abdalla, 2011). These newly created telecommunication companies offered the authoritarian state centralized control of the production and mass-dissemination of information and entertainment following an industrial/capitalist logic (Abdalla, 2011). In sum, Brazil simultaneously produced a void of cultural diversity occasionally thwarted by marginalized artists and intellectuals, as well as a growing and technically sophisticated industry of communication and entertainment.

During that time, non-white bodies gained wider visibility in a number of scenarios: Afro-Brazilian women were most widely seen as "professional mulatas," samba dancers who performed in G-string bikinis, feathered headdresses, and high heels (e.g. Sargentelli's *Show de Mulatas*). They danced at casinos, nightclubs, carnival parades, and on televised programs. Secondly, you could find them in Chanchadas films. These were burlesque comedies inspired by Hollywood musicals, yet motivated by Rio de Janeiro's music industry and its release of carnival songs every year. Finally, black and mulatto men also starred as footballers. Their popularity increased exponentially after Brazil won the World Cup championships in Chile (1962, broadcast in black and white) and in Mexico (1970, the first program to be broadcast in Technicolor in Brazil).

Hence, at the height of the military regime's socio-political censorship, football, samba, and carnival became significant contexts where both Afro-Brazilian bodies and their syncopated ways of moving were widely accepted and celebrated. On one level, Brazil celebrated black/mulatto (male) footballers such as Pelé and Garrincha as *exceptional* local heroes. Their athletic and virile performances contributed, perhaps like no other phenomena, towards authenticating the modern "democracy of races" ideology conceptualized in the 1930s. On another level, it also promoted local black/mulatto (female) samba dancers as widely available/disposable objects of desire. Following a *panem et circenses* logic, the movements of bodily syncopation that these entertainers articulate, their "hip(g)nosis," was explored to the fullest during the dictatorship. Ginga became a locally produced and cheaply disseminated opiate of the masses. The media also created the Brazilian that watches football, samba, and carnival on TV.

Generally speaking, Afro-Brazilian aesthetics and communities continued to be marginalized. However, the corporeal playfulness of footballers and the sensuality of samba dancers were packaged and exported as a trademark of Brazilianness. In the 1980s, the *Globo* TV station gained exclusive rights to broadcast Rio's samba school parades, transforming this popular manifestation into a made-in-Brazil commodity. In these videos, ginga played a particular script: like Sargentelli's shows, these televised parades privileged sexy (and shameless) mulatas shaking their hips for the camera wearing high heels and G-string bikinis.

President José Sarney's administration (1985–90) inaugurated a new era in the country. First, his administration founded an independent Ministry of Culture (1986) and the Fundação Palmares (1988), the first federal agency designed to promote the preservation, protection, and dissemination of Afro-Brazilian culture. Meanwhile, the state stepped aside and implemented a new piece of legislation that gave "the people" the right to intervene in the production of arts and culture in Brazil, remaining simply as a "facilitator" of this process. The so-called "Sarney Law" established a fiscal incentive geared towards the sponsorship of cultural products and services. Since then, mechanisms of tax deductions and waivers have dominated the cultural agenda of Brazil. As mentioned in Chapter 6, for instance, in 1989 Grupo Corpo signed a three-year agreement with Shell Brasil (later extended for ten years), which bound the company to produce a new evening length performance annually. In return, it received a maintenance sponsorship, which qualified under the Sarney Law.

In parallel to these state measures, the re-democratization of the country also led to a decentralized proliferation of political parties and non-governmental organizations, both of which sought to promote civil rights and racial equality, to restore a sense of self-esteem to marginalized communities, and to denounce all forms of exploitation and subjugation. In black cities such as Salvador, in particular, the gingado became an important ingredient of black pride discourse in Blocos Afros, Afoxés, and capoeira schools such as Grupo de Capoeira Angola Pelourinho. At times they joined forces with members of black political organizations (e.g. MNU, Brazil's Black Unified Movement) to negotiate their visibility in the wider social sphere.

After an initial period of euphoria, Fernando Collor de Mello's administration (1990–92) provoked a generalized destructuralization of the country's fragile emancipation. Yet, during this short mandate, the sociologist Sérgio Paulo Rouanet (Collor de Mello's nominated Secretary of Culture) departed from the Sarney Law to create a national program of cultural support on three fronts: a new Cultural Incentive Law, commonly known as the Rouanet Law (further discussed below in detail) and two others whose potential were never fully realized: the National Fund for Culture (FNC) and the Cultural and Artistic Investment Funds (Ficart) (MinC, 2010; Abdalla, 2011).

The Rouanet Law

In a nutshell, the initial Rouanet Law encouraged individual citizens and corporations to redirect six percent and four percent, respectively, of their taxable revenues to cultural projects. In return, they would receive tax incentives for what was spent. In the case of under-privileged sectors such as classical music and dance, however, the law granted 100 percent tax exemption. Contrary to the censorship laws instituted during the military regime (e.g. AI 5), under Brazil's new Cultural Incentive Law, feasibility is the only criteria employed to capitalize resources for cultural productions. Hence, artists, production companies, and cultural institutions have total autonomy over the content of their projects (i.e. no moral or political censorship). However, taxpayers (both individuals and companies) have the right to choose where, when, and how a portion of the taxes they owe the state (public money) is spent.

The fiscal incentives allocated to cultural projects represent a very small percentage of the total amount of incentives that the state subsidizes, especially when compared to similar laws that benefit sectors such as agriculture and sports. Nevertheless, today Brazil's Cultural Incentive Law is currently responsible for 90 percent of the total amount invested in the arts and culture in Brazil (MinC, 2010). Here is how we got there: during President Fernando Henrique Cardoso's administration (1995–2003), the original Rouanet Law went through a series of gradual transformations. As a result, by 2001, the revised piece of legislation expanded the 100 percent tax deduction criteria towards a wide range of cultural products and services.[8] In the end, this neo-liberal normatization of full sponsorship anchoring these cultural incentive laws meant that they could no longer be viewed simply as a "fiscal incentive" (Abdalla, 2011). Rather, legal mechanisms that grant 100 percent tax deduction merely legalized the transference of the role of managing public money allocated to culture into the hands of the private sector. Before I discuss in detail the benefits and the pitfalls of the incentives, let's take a moment to look at what it has produced and how it relates to the ginga aesthetic.

Brazilianness: which body is this? Which dance is that?

There are a number of contemporary artists and organizations whose movement research intersects with the ginga aesthetic or concepts and efforts associated with this system. Grupo de Rua de Niteroi (literally *Street Group of Niteroi*) is a dance company directed by the choreographer

Bruno Beltrão, which plays with ginga in its work. With *H2* (2005) and *H3* (2008), the group performs a crossover jump from hip hop competitions and re-invents itself as a contemporary dance company. By choosing to cast professional hip hop (male) dancers from Brazil whose regimens of bodily training range from capoeira to breakdance, the company has retained some aesthetic principles associated with ginga (e.g. polycentrism and polyrhythm). Yet, Beltrão's choreographies employ contemporary Western strategies to untangle and rewire essential elements shared by urban dances that proliferate across the black Atlantic world.

Other notable dance choreographers who have engaged with the ginga aesthetic in their works include Rui Moreira (a former dancer at Grupo Corpo), Luis de Abreu, Henrique Rodovalho, Mario Nascimento, Marcelo Evelin, Wagner Schwartz, and the duo Ângelo Madureira and Ana Catarina Vieira. Afro-Brazilian choreographers such as Moreira and de Abreu have sought to qualify and politicize aesthetic elements commonly associated with Afro-Brazilian heritage (including but not limited to the ginga aesthetic). In their own particular ways, these choreographers question the historical reduction and exploitation of their black dancing bodies as sexualized objects of desire. In *Samba do Criolo Doido* (*Samba of the Crazy Black Man*, 2005), Luis de Abreu stages a 20-minute controversial solo of agony, in which he implicates his own (black, male, queer) body in order to question stereotypical representations of blackness. Standing naked on stage or rather wearing nothing but a pair of high-heeled silver boots up to his knees, he performs a series of stomach contortions, framed by a Brazilian flag stretched in the background. The choreography is danced to (or perhaps against) a samba interpreted by the Brazilian diva Elsa Soares. As the choreography progresses, Soares screams repeatedly, with her characteristic high-pitched and ragged voice, "the cheapest meat in the market is the black meat" (*a carne mais barata do mercado é a carne negra*).

Solo performers such as Evelin and Schwartz have appropriated ginga in a more conceptual way. After living in Europe in the recent past, the idea of ginga arose symbolically in their reflections over their migrating bodies. In *Wagner Pina Miranda Xavier Ribot Le Schwartz Transobjeto* (2004), for instance, Schwartz unpacks and further problematizes static conceptualizations of and about (his) identity in the age of globalization. On the micro level, he gravitates towards the work of Hélio Oiticica and Lygia Clark, inasmuch as they "problematized the universality of things" (Schwartz, 2014).[9] On the macro level, however, he utilizes their concept of "cultural cannibalism" to "devour" and "recycle" both old clichés specific to Brazil and new tendencies articulated in the contemporary dance scene in Europe. On his website's blog, Schwartz talks about his conceptual and minimalist solo:

> In *Wagner Pina Miranda Xavier Ribot Le Schwartz Transobjeto* there is no La Ribot, Pina Bausch, Xavier Le Roy, Carmen Miranda, samba, favela, ginga, colored postcard, Central Plateau [the location of Brazil's capital]. But [there is] a thought that took shape at the margins of La Ribot, Pina Bausch, Xavier Le Roy, Carmen Miranda, samba, favela, ginga, colored postcard, Central Plateau. (Schwartz, 2006)

Closely related to Schwartz's "inquietudes" is the work of PIP, a contemporary dance company founded by the dancer, choreographer, and art director Carmen Jorge. In *3 Mg – Gingaestética* (2005) and *Barraco* (2006), Jorge experiments physically with Paola Berenstein-Jacques' understanding of the fragment, the labyrinth, and the rhizome. They are the constituting elements of her "aesthetic of ginga" (see Chapter 1). PIP's creative projects borrow elements from Oiticica's interactive installations and body-events and Berenstein-Jacques' concept of movement-space. On PIP's website (www.pip.art.br), one finds that in *3 Mg – Gingaestética*:

> The *Ginga* appears in the body and in the space through the dynamics that keeps the movement of the images all the time ephemeral. As a metaphor to [sic] the labyrinth that advances always maintaining the possibility of transformation, where the performers' presence and perspective are the generating elements of instability in the scene.

PIP's creations fall under the umbrella of "integrated arts." In *Barraco*, for instance, Jorge explores the favelas' "makeshift logic" and its underlying aesthetic principles. Briefly, dancers interact with a high-tech/low-tech multimedia installation containing 29 TV sets, 40 plants, and two groupings of beach chairs (designated for the audience). Binding the piece together, dancers improvise a choreography whose dynamics of movement are anchored in principles such as "fragmentation," "risk," and "precariousness." In doing so, their movement interactions make loose references to gambiarras or improvised bricolages and "make-dos" developed out of necessity (see the Introduction). As I see it, these improvised choices may be understood as a process of "self-editing" produced by bodies that are at the same time alert and overflowing.

Jorge's "self-editing" process relies on the understanding that, similar to what we witness in Grupo Corpo's *Breu* (2007), in hostile environments, people are constantly improvising solutions or "making do" to the extent that risk and precariousness becomes a way of life. Unlike Pederneiras' set scores, however, Jorge pushes her dancers-collaborators to improvise gingas-gambiarras as a way of thinking about and moving across the scenic space, that is, to articulate bricolages and makeshifts through interaction; to think on their feet and take risks on the fly (see the Introduction). Most importantly, PIP's creative research has incorporated elements of the aesthetic of the favelas into both its form (e.g. syncopated articulation of body and space) and content/context (e.g. risk, makeshift, gambiarra).

On the other side of the spectrum, the Japanese-Brazilian designer Jum Nakao has also appropriated the idea of gambiarra in his design projects, further expanding this already broad articulation of ginga, makeshift, and improvisation. For Nakao, an artist who defines himself as a Japanese product assembled in Brazil, Brazilianness means the ability to "grab the whole world, to amalgamate and give it back as an unique form, filled with gambiarras, with ginga." In 2012, the designer's name generated much controversy in the media when it was announced that he was designing the costumes for Brazil's eight-minute performance at the end of the Summer Olympics closing ceremony in London. During this made-for-TV spectacle, the Olympic torch would be passed on to the mayor of Rio de Janeiro, the city hosting the next sporting event in 2016. The performance of Brazil's delegation was co-directed by the theatre director Daniela Thomas and the filmmaker Cao Guimarães, who were expected to deliver a technically impeccable piece of live entertainment for a mainstream audience.

In the weeks leading up to this much-anticipated live event, there was a lot of buzz raised across Brazil's printed and virtual media, starting with those who questioned Nakao's ability to "truthfully" represent Brazilianness internationally. By the same token, others feared that the Brazilian performance was going to boil down to a patriotic projection of Rio de Janeiro as an exotic, happy-go-lucky "melting pot" of cultural clichés, where the "flag" of racial democracy is raised, without any critical considerations. During a debate led by choreographers Marcelo Evelin and Dani Lima at Núcleo do Dirceu, for instance, some of these edgy and restless inquiries came to the surface. In a summary of the said discussion, published on Nucleo do Dirceu's website, in a forum entitled "Coleção Brasilidade apresenta: o gesto" ("Collection Brazilianness Presents: The Gesture"),[10] it is stated that:

> The multicultural discourse is another extreme, the blasé of "folks, Brazil is the world." The cliché of mixture. In the Olympics there

> will be Africa and Japan in the bring-down-the-house carnival style. This is our Brazilian identity, the idea that we are many to a point that we can be "almost anyone, anything. Hail to multi-ethnic Brazil, miscegenated, plural!!!" ... And now that we are being summoned to show what we understand by Brazilianness to the world, which body is this? Which dance is that? (Author's translation)[11]

In many ways, they were right. Though impeccable, the representation of "Brazilianness" in the closing ceremony may be summarized as a visual hurricane of equally dazzling and essentializing figures. At the same time, on the morning after the closing ceremony, Jum Nakao published a long note on his Facebook status entitled "Anthropophagy of Our Own Clichés." Making a clear reference to Oswald de Andrade's 1928 *Manifesto Antropofágico,* Nakao states that: "What most defines Brazilianness to me is this anthropophagic culture." He then addresses the connection between his playful act of mixing things and the idea of ginga-gambiarra:

> Departing from a script that was sent to us, we designed all the costumes. We tried to valorize popular manifestations through *gambiarra* as an artistic language resource. The *gambiarra* or improvisation is one of the marks that characterize the artistic production in Brazil, with the artistic assimilation of street and popular procedures and aesthetics. (Author's translation)[12]

As the examples above exemplify, the tug of war over what it means to be Brazilian in general and how the ginga aesthetic factors into this equation in particular is still ongoing. Several artists have found, within their distinct spheres and proposals, innovative ways to appropriate and take liberties with this syncopated logic, whether they call it "the aesthetic of favelas," "gambiarra," or simply "swagger." This aesthetic system enables them to question and rearrange both local and foreign materials, dismantle some ties and strengthen others in a rhizome-like fashion, juxtapose unpredictable alliances and fragmentations, or articulate dissonant utterances.

However, there are a number of artists and intellectuals who fear that the carnivalesque understanding of Brazil as a homogeneous melting pot may obfuscate social inequality and cultural discrepancies. Others such as Lima and Evelin point out that this celebration of "harmonic mixtures" leaves little room for critical inquiries towards diversity in terms of both body types and dance forms. After all, is it possible, as Nakao proposes in relation to his Facebook status, to "culturally cannibalize our own clichés and values" without questioning the (bio/geo)politics inherent to contexts in which these images circulate in the media?

Despite the re-democratization of the country, its economic boom, and education reform, every new summer Globeleza (see Chapter 3) re-emerges, anticipating Globo's televised broadcasting of Rio de Janeiro's samba school parades. In Rio de Janeiro's favelas, the macho overtones of the local "funk" scene (*funk carioca*) and its music videos continue to mash together the same old formula of exploitation that connects the ginga aesthetic and blackness to feminized submission and sexual immorality. These two examples, amongst others, are clear signs that the pride-and-shame conundrum attached to this way of moving is yet to be solved or overcome.

Behind the scenes: the Rouanet Law in numbers

Over the last 20 years, the Rouanet Law brought about a number of tangible benefits. For the sake of clarity, I will refer to Brazil's oil company Petrobrás as an example of how the system works in practice. Regarded as the sixth largest energy company in the world and the most profitable company in the country, in the last decade Petrobrás was also put in charge of directly managing 20 percent of Brazil's cultural and artistic production. Even areas that do not normally appeal to the masses, such as experimental dance and classical music, have seen benefits from these initiatives.

Thanks to the contributions by large corporations like Petrobrás, the landscape of contemporary dance in Brazil saw a boom of organizations (e.g. groups, companies, and collectives). New productions circulate both nationally and internationally. Petrobrás has channeled a substantial amount of money towards the production of dance projects from classical to cutting-edge genres. At its higher end, Petrobrás sponsors the maintenance of two well-established dance companies from Brazil, Grupo Corpo and Deborah Colker . Each of these two companies receives around four million reais (nearly $2 million in 2014) from Petrobrás each year to cover costs relating to its "maintenance," which may include the production and circulation of dance performances as well as the salary of its artistic and technical staff. Since 2005, it has also sponsored international dance festivals, such as Rio de Janeiro's *Festival Panorama de Dança* (founded in 1991), Belo Horizonte's *Fórum Internacional de Dança* (founded in 1996), and Fortaleza's *Bienal de Dança* (founded in 1997). Finally Brazil's oil company has also funded, via the

Rouanet Law, the development of specialized publications, periodicals, blogs, archives, documentaries, etc. Created in 2003 and "sponsored" by Petrobrás since 2006, the online magazine *idança* (www.idanca.net) is today one of the most widely read sources of information about dance in the Portuguese language.

There are several pressing factors associated with these incentive laws that have sparked a number of heated discussions and political mobilizations. First, when the Brazilian government implemented the Rouanet Law in the 1990s, it sold the idea of investing in culture to the private sector as a communication and marketing strategy to improve the image of their brand names. Meanwhile, this neo-liberal approach transferred to the market the responsibility for managing the production and dissemination of *cultural products* in particular, thus contributing to the commodification of culture (Abdalla, 2011). Secondly, by granting the private sector the right to manage public funds (waived taxes), the state also removed off the table the corporations' responsibility to pour their own dividends into cultural projects (Abdalla, 2011). This reductionist understanding of culture as a commodity that may add kudos to the branding of corporations is governed by the market economy. Hence, it tends to valorize products connected to the industry of entertainment rather than projects that might be created by conscious citizens and/or offer critical reflections of their reality. Thirdly, despite the increase in the overall quantity, the quality of artistic projects often remains below expectations. Currently, there are no mechanisms in place to regulate the artistic merit or cultural relevance of a project funded through these incentive laws.

The economic growth of the country during Luiz Inácio Lula da Silva's administration (2003–10) led to an even wider explosion of cultural projects, many of which were disproportionally large in scale. This system continues to perceive and treat the most varied artistic projects as cultural commodities. Subsequently, in the last decade, the aesthetic inequalities and ethical disparities that resulted from this system have fueled a growing number of complaints in Brazil. In order to critically analyze how the market has regulated and shaped up the cultural production in the country, below I will follow the money trail.

According to the Ministry of Culture (MinC, 2010), since its creation in the early 1990s, 80 percent of all incentives were poured into the southeast region of Brazil. Rio de Janeiro and São Paulo received 50 percent of that sum. At the same time, the Ministry of Culture's report recognizes that most Brazilians have never attended an art exhibition or dance concert and that 92 percent of all municipalities in Brazil still lack museums, theaters, or cinema venues (MinC, 2010).

Zooming in a little closer, one finds that nearly half of all public investments in the last two decades via the Rouanet Law have been allocated to four particular areas: classical music (9 percent); books (10 percent); theater (11 percent); and the integrated arts (14 percent), an umbrella term that encompasses performing arts projects that intersect, but are not limited to, theater, dance, circus, music, and the visual arts. Hence, in the last two decades, the market has privileged the creation and circulation of cultural *products* (e.g. art books, music CDs and DVDs, theatrical performances, dance concerts, etc.) with relatively short turnaround times rather than investing in long-term cultural *research* and *processes*, whose results may or may not lead to tangible commodities. With the exception of Petrobrás' maintenance of Grupo Corpo and Deborah Colker, dance companies receive between 500,000 and 600,000 reais (up to $300,000 in 2014) annually for a maximum of three years. Hence, in reality, if one considers that a professional dance company should offer a formal contract to its working artists and staff, the sponsorship is insufficient (Katz, 2013). As a result, most companies in the country maintain an "informal" hiring policy, where dancers are invariably penalized or put at risk without basic work rights (e.g. health insurance, sick pay, or worker's compensation in the event of injury).

The Intangible Heritage Law

The Ministry of Culture's report also revealed that a pool of 30 "minor segments," which encompasses an array of heterogeneous categories (e.g. library, archive, history, photography, circus, opera, folklore, popular culture, Afro-Brazilian culture, and Indigenous culture) have collectively received only 14 percent of all tax-break incentives (MinC, 2010). Beyond the private sector's unequal distribution of public funds, I find it particularly intriguing that "Afro-Brazilian culture" and "Indigenous culture" are accounted for, and thus understood, outside broad categories of cultural production such as theater, dance, music, and integrated arts (and even popular culture and folklore), as if the products and process articulated around these ethno-cultural heritages could not be measured or qualified under standard (i.e. Western/colonial) criteria of art. In the end, one must ask: what kinds of combinations and segregations have this arbitrary set of categories established under the Rouanet Law? What kinds of biopolitics inform these discriminations and how can they be overcome? What effects do they generate?

In order to answer this new set of questions, let's go back to August 16, 2014. At first glance, it seemed as if the state had come to recognize the importance of supporting different cultural forms of expressions, practices, and procedures. In fact, all the cultural events I was hoping to attend that day benefited from the Cultural Incentive Law. Nevertheless, upon a closer look, it is possible to identify two distinct patterns, that is, two different ways of promoting or celebrating cultural goods in Brazil. On one level, there are the (capitalized) Arts and Culture, whose (commoditized) products fit within Western traditions, processes, and concerns, and whose work (and effort) is often shaped by authorship, formal training, and/or (perceived) ability to "evolve" and "innovate." Coincidently or not, the categories that most benefited from the Cultural Incentive Law in the past 20 years, i.e. art books, theater, music, and the integrated arts, all fulfill this description. On the other level, other categories such as "Afro-Brazilian culture," as well as "Indigenous culture," "popular culture," and "folklore," have been historically valued for their social role rather than their artistic merit. Also contrary to the "classical" or "innovative" cultural commodities outlined above, these "traditional" forms of expression are shared by the community rather than "invented" by a person. More often than not, they are maintained through informal training centered on the (master-disciple) guild system and on oral tradition. As the statistics in the Ministry of Culture's report confirm, these "traditional" practices, religious celebrations, and trades have presented little appeal to the market and thus have received little financial support via the Rouanet Law.

In parallel to these market-driven measures, in 2000, Congress approved the Intangible Cultural Heritage Law (Lei 3.551/00) under which to properly *catalogue* and *safeguard* these "endangered" cultures. Ironically, the entertainment industry and the means of mass communication, one of the biggest enemies of non-Western cultural heritage, often benefit from the market-obeying cultural incentive laws such as the Rouanet Law. Despite the discrepancies between the implementation of these two kinds of public policies, the National Program of Intangible Heritage of Brazil's National Institute of Historical and Artistic Heritage (*Instituto do Patrimônio Histórico e Artístico Nacional* (IPHAN)) has developed groundbreaking research and has published dossiers with the intention to archive the relevance of Brazil's diverse cultural influences. Contrary to the Rouanet Law, IPHAN's certificates are not attached to any financial benefit, nor do they help cultural agents and organizations to mobilize donations and/or sponsorships from the private sector.

From the catalogued and safeguarded practices centered on the ginga aesthetic, it is worthwhile noting: the samba de roda of Bahia's countryside (2004); the trade of baianas who sell acarajé (2005); the southeast's jongo (2005), Maranhão's tambor de crioula (2007); Pernanbuco's frevo (2007); and the capoeira circles and the trade of capoeira mestres (2008). Today, the practices, rituals, and trades included in this "intangible heritage collection" are recognized as building blocks of Brazil's cultural past and its current way of living, acting, and creating. These professions and expressions are no longer rejected or persecuted as was once the case. But the problem of how to re-integrate these aesthetic knowledges and ways of knowing within Brazil's mainstream systems of education and cultural production is yet to be solved. So far, these practices have survived *in spite of* the Brazilian government, the market, the industries of entertainment, and the elites.

The tradition of samba circles practiced in the countryside of Bahia, the so-called *samba de roda do recôncavo baiano*, was one of the first Afro-Brazilian manifestations to be recognized as an immaterial patrimony of Brazilian culture in 2004. Although it would be erroneous to identify Bahia or its recôncavo as the "birthplace" of this way of dancing, there is a vast amount of information available about the samba circles practiced back in the nineteenth century in that once economically thriving area. IPHAN's research efforts and its published dossier have helped researchers, dancers, and enthusiasts to better understand how, despite adverse conditions, Afro-Brazilian communities continue to maintain their cultural memory alive. For Carlos Sandroni (2004), the ethnomusicologist who organized the published dossier, the importance of the samba de roda, as practiced in that region, transcends its historical character inasmuch as it is an intrinsic part of the daily practice of men and women in Bahia. In doing so, the dossier he organized indirectly recognizes samba dancers and musicians as agents responsible for both the preservation of their aesthetic knowledges and the re-creation of their (ethno-cultural) processes of identification.

Similarly, on July 15, 2008, IPHAN recognized the practice of capoeira as an immaterial patrimony of Brazilian culture. The recognition was organized into two subcategories: (1) the "trade of the mestres" (*Ofício dos mestres*) indexed within the *The Book of Archived Wisdoms* (*Livro do Registro dos Saberes*); and (2) the "capoeira circle" (*roda de capoeira*) indexed within the *Book of Archived Forms of Expressions* (*Livro do Registro das Formas de Expressão*). In the end, what remains intriguing is the fact that the archived description of these "immaterial cultures" recognizes the role of the agents historically involved in these cultural manifestations, but makes no provision to give them any (financial) incentives to keep their practice alive. Samba dancers, capoeira players,

Candomblé priestesses, and the street food vendors constitute some of the figures that Brazilians associate most frequently with Bahia and, subsequently, with their Afro-Brazilian "roots." However, since these professions are not regulated by the state, these citizens are yet to receive any tangible benefit that could safeguard their lives and/or support vital necessities for them such as health insurance or retirement plans. As the capoeira Mestre Pastinha once said, the capoeira is doing well; the capoeira mestres are the ones in need of help. Like contemporary dancers and choreographers working with "cultural products," the agents of traditional Afro-Brazilian "cultural manifestations" constitute a population at risk.

Since the 1990s, an increasing number of capoeira players have chosen to emigrate to other countries or travel abroad on a regular basis in search of social recognition and mobility. In fact, today some of the most distinguished mestres and professors of capoeira either live in or constantly travel to Europe, the US, and other so-called "developed countries." Yet, despite the increased visibility and celebration of capoeira abroad, within Brazil, it remains practically impossible to earn a living as a capoeira mestre. As a result, capoeira events and festivals, such as the *Agosto da Capoeira*, offer a productive way to sparsely finance the well-being of these invaluable repositories of living culture. Nevertheless, like the "sponsorship" of concert dance via the Rouanet Law, these investments are insufficient.

Dance and education

In parallel to all of this, throughout the 1990s, Cardoso's administration implemented a series of policies that facilitated the development of new university programs in the areas of art production, education, and management, including the visual arts, theater, music, and dance. Hence, if from the 1950s to the 1980s there was only one dance program in the country (UFBA), by 1999, that number had increased to ten programs. Beyond these "official" dance programs, other institutions developed interdisciplinary programs that have nonetheless resulted in projects related to the field of critical dance studies. A good example is the undergraduate program in Communication of Bodily Arts (*Comunicação das Artes do Corpo*) created in 1999 at the PUC-SP. During Lula's administration (2003–10), new public policies incentivized the creation of new public universities with dance programs. Today there are over 40 bachelor's programs in dance spread across 33 institutions and one MA program in dance studies at UFBA.

In 2003, Lula's administration instituted Law 10.639, which mandates the inclusion of African and Afro-Brazilian culture into primary and secondary school teachings. Along with affirmative actions, this piece of legislation has not only promoted a massive questioning of the curricula of Brazilian primary and secondary schools (especially in relation to performing arts and literature), but has also generated a demand for courses and programs in higher education that may produce and prepare educators to fulfill such demands. However, from the discussions during the dance educators' conference at UFBA in 2014, primary and secondary school teachers, especially those hired under the public school system, expressed other opinions. Many are facing a number of structural difficulties in implementing a diverse curriculum that also includes African and Afro-Brazilian arts and culture. Often, as educators at the conference confessed, Afro-Brazilian forms and practices, and their underlying aesthetic knowledges, are only given proper attention during civic dates such as "Folklore Day" (August 22) and "Black Consciousness Day" (November 20), when dance teachers are coerced to "put on a show" for the school. This is not to say that these measures are inefficient, but rather that the complex process of the education of the relevance of local arts and cultures connected to the African (and Amerindian) heritages in Brazil to the development of the dignity and self-worth of all its citizens will take more than a few laws and decrees to be fully realized.

In 2010, the Rouanet Law was once again modified, this time in response to the complaints and concerns expressed by different segments of the society. Briefly, this new set of reforms offers some improvements, such as the reduction of the percentage that may be deducted from taxes and the introduction of the "culture card," a type of food stamp for culture. Nonetheless, the issues are far from being resolved. Also in 2010, the Ministry of Culture established the National Plan of Culture and the National Plan of Dance. These "plans" present series of long-term public policy strategies and measures, which take into consideration: (a) the role of the state in assisting the production of arts and culture and the participation of society in these decisions; (b) the protection and promotion of diversity; (c) access to cultural goods; and (d) socio-economic development and sustainability.

Capoeira, permacultura, and care of the self

Before you put this book down, let me take you back to the Kilombo I told you about at the beginning of this chapter. Like many other capoeira players, in the 1990s, Mestre Cobra Mansa moved to the US in search for better opportunities and, shortly afterwards, opened an independent capoeira school in Washington DC, which later came to be known as the ICAF. Ten years ago, Mestre Cobra Mansa returned to Brazil and began to cultivate a revolutionary idea, which sprang from his newfound interest in renewable and sustainable modes of living. It has since attracted a variety of people interested in implementing principles of permaculture. At Mestre Cobra Mansa's Kilombo, individuals learn much more than sustainable agriculture, natural building, agroforestry, rainforest harvesting, and apiculture. At the heart of this multifaceted project is capoeira angola, practiced every morning at dawn and all day long during its semi-annual events, such as the *Permangola.* Though located in a remote location, the Kilombo's cultural events have attracted an array of professionals from all over the world, including engineers, architects, environmentalists, agriculturalists, gardeners, herbalists, cooks, healers, artists, musicians, dancers, and, of course, capoeira players. Over time, they all come into contact with capoeira's aesthetic and philosophical knowledge as well as its underlying logic of organization and production, the ginga aesthetic.

In the near future, I intend to further examine how this cultural and ecological center combines this collection of local and global approaches (i.e. ways of planting, harvesting, cooking, moving, and playing music) in its quest to recuperate-cum-invent sustainable ways to take care of oneself, others, and the surroundings they inhabit. As I have now just began to understand, at the Kilombo, the ginga aesthetic is more than the structuring element within the practice of capoeira angola. At times, it is extended as a common denominator which the people working at that farm employ to orient themselves, both inside and outside the capoeira circle. What remains to be further investigated and examined is how this embodied logic, which is practiced daily, interacts with and germinates in an ecosystem grounded in solidarity and sustainability rather than risk and precariousness (i.e. favelas), where it had historically been cultivated in Brazil.

Inspired by ideas such as technologies of the self (Foucault, 1988),[13] as well as movable territories, corpomedia, and corporal episteme (see the Introduction), the hypothesis that I would like to further investigate may be summarized as follows: has the ginga aesthetic functioned as a significant element within this alternative way of being-in-the-world cultivated and promoted at the Kilombo Tenondé? And, if so, how will the cultivation of this Africanist system of organization and production (shared by all its participants) inform the ecological and cultural premises of a project grounded in practices that promote solidarity and sustainability? And what effects will it produce over time? Finally, will this international project transplanted back to Brazil contribute towards overcoming the country's pride-and-shame conundrum when explored to its full capacity?

Notes

Introduction: Choreographing Ideas

1. Gingado is the past participle of the Portuguese verb "gingar" (to sway or to swing). In this case, it acts as a noun. While as a noun "ginga" means "the sway," gingado may be translated as "the swayed action." Although they may be employed as synonyms, during the first part of the twentieth century, the term "gingado" was used more frequently.
2. Livia Barbosa (1992) presents a socio-anthropologic analysis of the particular "way" (*jeito, jeitinho*) in which people (are expected to) act and behave socially in Brazil. For Barbosa, the inclination towards finding an alternative or personal way to slip through what might otherwise look like a "dead end" or "red tape" is a Brazilian institution. Furthermore, according to Roberto DaMatta (1987, 1997), this "Brazilian way" of improvising innovative solutions to unforeseen scenarios may be understood as an acquired behavior that has been carefully constructed and continuously performed. This way of making do, he points out, has given unified meaning to a diversified (social) body within the discourse of national identification in Brazil.
3. As a figure of speech, the terms "ginga," "jeitinho," and "jogo-de-cintura" may be used interchangeably. In Brazilian Portuguese, they evoke a metaphoric juggling of ideas or propositions employed as a strategy of negotiation or mental bargain in society. They may also reference one's personalized way of overcoming hindrance, red tape, or prohibition, or, rather, a way of "fixing" a problem or a situation.
4. Ricardo Rosas writes that: "The Portuguese word gambiarra immediately conjures images of the clandestine electricity hook-ups so often seen in slums and shantytowns, and this is precisely the first sense of the word as defined by the dictionary Houaiss. However, gambiarra also means something far more akin to the English term 'make-shift,' referring to any improvisation of an expedient substitute when other means fail or are not available. In other words, 'making do'. In Brazil, the term carries an especially strong cultural weight, being used to define any quick-fix solution made with whatever happens to be at hand. This sense of the term has not been lost on the art scene, and features in various creations in the field of the visual arts. Indeed, it is from this harvest in particular that we can identify further and revealing concepts behind the gambiarra and its symbolic/cultural meaning" (Rosas, 2006, pp. 37–8).
5. Wikipedia lists ten different meanings for the word "ginga." In the UK, ginga (or ginger) is a derogatory term used to describe people with red hair. In Japanese, however, ginga means *galaxy*, and in that country this term may refer to a satellite, a train, a bomber aircraft, an anime series, two fictional characters, and the title of a song.
6. Pioneers in the field include Mário de Andrade (1893–1945), Câmara Cascudo (1898–1986), Arthur Ramos (1903–1949), Cecília Meirelles (1901–1964), and Edson Carneiro (1912–1972).

7. Informing this book are the ideas planted by Katz (1989), Browning (1995), Araujo (1996), Cabral (1996), Sodré (1998), Vianna (1999), Bruhns (2000), Lopes (2003b), and Chasteen (2004).
8. Examples include *The Samba Circles of Bahia's Countryside* (*Samba de Roda do Reconcavo Baiano* (2004), *The Trade of the Baiana do Acarajé* (2005) and *The Roots of Samba from Rio de Janeiro* (*Matrizes do Samba no Rio de Janeiro* (2005).
9. From the publications that have significantly broadened the understanding of this martial art, it is worthwhile noting the works of Tavares (1984, 1998), Thompson (1988), Dossar (1992, 1994), Pinto (1991), Vieira (1991), Lewis (1992), Reis (1993, 1997), Desch-Obi (1994, 2000, 2008), Browning (1995), Soares (1994, 2004), Downey (1998, 2005), Assunção (2005), Abib (2004), Merrell (2005), Desch-Obi (2008), and Talmon-Chvaicer (2008).
10. Some of my strong influences include Foster (1986, 1995, 1998), Savigliano (1995), Daly (1995), Albright (1997), Banes (1998), Martin (1998), Burt (1998, 2007), Desmond (2001), Albright and Gere (2003), Manning (2004, 2006), Lepecki (2006), Shea Murphy (2007), O'Shea (2007, 2010), and Kraut (2008).
11. Greiner and Katz are professors in the Communication of Bodily Art program at the Pontifícia Universidade Católica de São Paulo (PUC-SP).
12. For Katz and Greiner, contamination (of knowledges, points of view, etc.) is the norm. Hence, ethno-cultural purity or purification is nearly impossible to maintain or achieve once people are always "contaminating" and being "contaminated" by one another.
13. For Flusser, *gestures* are the first and most basic code or the departure point for the constructions of mental processes. He further proposes that codified texts (e.g. images, books, or videos) are nothing but materialized gestures.
14. According to Foster, "choreography, whether created by individual or collective agencies, improvised or designated in advance, stands apart from any performance of it as the overarching score or plan that evidences a theory of embodiment. This plan or framework of decisions that implements a set of representational strategies is what endures as that which is augmented, enriched, or repressed in any given performance" (1998, pp. 16–17).
15. The expression "recuperation-cum-invention" in my text draws attention to the ways in which Africans and their descendants have not only preserved or restored situated knowledges (Mignolo, 2002), judgments of taste (Bourdieu, 1984), and corporeal epistemes (Foster, 2011) connected to their ethno-cultural heritage in the New World. The cultivation of non-hegemonic ways of thinking and acting such as the ginga aesthetic has enabled disenfranchised individuals and communities to both *recuperate* their sense of self-esteem and dignity and *invent* renewed identities that connect blackness to concepts such as grace and pride. Furthermore, the set or improvised choreographies they articulate of themselves as individuals or communities have *affected* their power of acting and thinking, through multiple processes of mutual "contamination" and "transculturation." In the end, I argue, these active processes have empowered them to resist or "fight against" practices and discourses of subjugation.
16. See Rose (1996), DeFrantz (2004a, 2004b), and Daniel (2005).

17. Santiago Castro-Gómez affirms that: "According to the taxonomy recently proposed by John Beverley, the field of cultural studies in the past ten years can be divided into four distinct albeit not complementary categories." The Walter Mignolo and the "Coloniality of Power" group, as he puts it, "includes Edgardo Langer, Aníbal Quijano, Enrique Dussel, Catherine Walsh, Javier Sanjinés, Fernando Coronil, Oscar Guardiola, Ramón Grosfoguel, Freya Schiwy, and Nelso Maldonado, along with myself" (Castro-Gómez, 2008, pp. 259–60).
18. See Browning (1998).
19. In Brazil, these ideas were disseminated by intellectuals such as the French diplomat Joseph Arthur, Comte de Gobineau (1816–82) and the Brazilian physician Raimundo Nina Rodrigues (1862–1906).
20. In the classic novel *Memórias Póstumas de Brás Cubas* (1881), cited in the epitaph of this chapter, the narrator explains that the "acrobatic idea" he had "was nothing less than the invention of a sublime remedy, an anti-hypochondriacal poultice, destined to alleviate our human melancholy" (Machado de Assis, 2006, pp. 5–7, translated by the author).
21. In groundbreaking works such as *Casa-Grande & Senzala* (*The Masters and the Slaves*, 1933) and *Sobrados e Mucambos* (*The Mansions and the Shanties*, 1936), Freyre deviates from the long-lived anxiety regarding Brazil's *mixed-race population* and its past of enslavement as sources of shame. Instead, his scholarship celebrates the idealization of Brazil's *sublimated and homogenized culture* as a source of pride.
22. The 1950 World Cup was the first tournament held after the Second World War. Originally scheduled to take place in 1946, it was the first World Cup attended by the British team.
23. The "Revolution of 1930" glorified the value of (productive) work, but also developed new tactics to eradicate "roguery" and "cunningness" practiced in urban centers by members of small gangs and isolated pick-pocket criminals, largely composed of unemployed blacks and mulattos.
24. President Getúlio Vargas' populist dictatorship was modeled after A.O. Salasar's *Estado Novo* (1933–68), a Portuguese authoritarian regime that enforced nationalist and Catholic values on its citizens. Exalting the five centuries of Portuguese maritime expansion, the regime resisted the process of decolonization that accelerated after the Second World War.

1 Decoding the Ginga Aesthetic

1. "Ginga é um jeito de não levar a vida muito a sério e de encarar os problemas com um jogo de cintura, pés e calcanhares. Há 505 anos que os brasileiros vêm gingando pela vida - e podem recomendar isso para todos." Cited in the documentary *Ginga: The Soul of Brazilian Football* (2005).
2. The *Ginga: The Soul of the Brazilian Football* documentary was written and directed by Tocha Alves, Hank Levine, and Marcelo Machado (2005), all of whom work for the O2 Filmes production company. It was co-produced by Nike, Wieden+Kennedy Entertainment (an American advertisement agency and longtime collaborator with Nike) and two partners at O2 Filmes: the

German Hank Levine (*Waist Land*) and the Brazilian Fernando Meirelles (*City of God, Constant Gardner, Blindness*).

3. Sandroni's use of the terms "cometricity" and "contrametricity" draws on the scholarship of the ethnomusicologist and composer Mieczyslaw Kollnski. For him, the syncope present in Brazilian music is nothing more than the way in which this frictional interplay between *cometricity* and *contratermetricity*, characteristic of African music, is represented in Western musical notation (2001, p. 26).
4. For Gottschild (1998), the Africanist principles include: *polycentric and polyrhythmic body, embraced conflict, high-affect juxtaposition, ephebism* and *the aesthetic of the cool.*
5. Drewal and Pemberton (1989) proposes that while the cold principle *iwa* relates to one's essential nature as a divine force, the hot principle *asé* ignites the world and may be translated as "life force" or "the power to make things happen."
6. Maxixe is black-creole ballroom style that gained visibility during the final decades of the nineteenth century in the outskirts of Rio de Janeiro.
7. For Sodré, samba's "magnetic force, compulsive even, comes from the impulse (provoked by the rhythmic lacuna) to fill in the absence of time with the dynamic of movement in space" (1998, p. 11).
8. Regarding Western view on play, see Huizinga (1955), Marcuse (1974), and Bataille (1986).
9. For the Nagos and Jejes, as the Yorubas are known in Brazil, the cosmos is imagined as a spherical gourd or as a circular divination tray (*Opon Ifá*). The mythic scenes or daily concerns depicted on such calabashes and trays represent community activities. The Opon Ifá functions as a field of (metaphysical) play, chance, and improvisation. In the process of divination, the priest first inscribes the empty circle with two perpendicular lines (crossroads) that connect the two halves of the cosmos: the physical/tangible world of living (*ayé*) and the invisible/spiritual otherworld (*orun*) of ancestors and supernatural beings. The priest invokes Exu, the trickster *orixá*, to siphon the performative force (*asé or axé*) onto such three-dimensional space and infuses the cowry shells with the powers to play open-ended metaphors (*ngangas*). Finally, the priest casts shells repeatedly on the board to foresee and interpret the future.
10. For Henry Drewal: "Like 'signifyin(g),' the northern African-American tradition of subtle (and not so subtle) commentary, satire, and cultural critique … the siré of Afro-Brazilian carnival performances uses the transformations, the parade of masking, as a primary mode of expression – masking that disguises as it disclosures, hides as it reveals, formulates as it transforms" (Henry Drewal, 1999, p. 164). Much like the Signifying Monkey tales of African-American vernacular culture and the related concept of signifyin(g) in arts and literature (Gates, 1988; Caponi-Tabery, 1999), the playful discourses enacted within Afro-Brazilian forms such as samba dance and capoeira evoke serious dialogues of playfulness and trickery. In particular, the fake and flickering processes enacted in these practices of cultural resistance, and the indefiniteness they provoke, function as fractured/dissonant articulations and figurations, which always already differ and defer fixed identifications.

11. www.softwarepublico.gov.br/ver-comunidade?community_id=1101545 (date accessed May 14, 2015).
12. Pontifícia Universidade Católica do Rio de Janeiro e Universidade Federal da Paraiba.
13. www.ginga.org.br (date accessed April 13, 2015).
14. Flusser's monograph was originally written in German shortly after he returned to Europe in the 1970s. It was eventually published posthumously under the title *Brasilien oder die Suche nach dem neuen Menschen: Für eine Phänomenologie der Unterentwicklung* (*Brazil or the Search for a New Man: For a Phenomenology of Underdevelopment*) (Aachen: Bollmann Verlag, 1994). *Gestures* was published in English in 2014.
15. Although Flusser uses the word "Brazilian" in his text, he acknowledges that this "structured behavior" or disposition derives from African heritage.
16. Unlike Western religions, which see the body only as a vessel for the soul, Flusser proposes that Brazilians have incorporated, in particular, the highly sophisticated and complex ability to produce, experience, transmit, and disrupt cultural meaning through the articulation of polyrhythmic actions or the African syncope.
17. Berenstein-Jacques' conceptualization of the rhizome departs from Deleuze and Guattari's theoretical conceptualization of the term in relation to the idea of "multiplicities." As an architectural concept, the rhizome privileges the horizontal, lateral, decentralized, polyrhythmic, and moveable organization against the stability of the vertical "tree-root" growth system.
18. In 2003, Lia Rodrigues Compania de Danças moved to Nova Holanda, a neighborhood inside the Maré Complex, today one of the largest favelas in Rio de Janeiro.
19. Video interview published in 2010 on the Panorama Festival's YouTube channel: www.youtube.com/watch?v=eMmFvCXdx3g (date accessed April 13, 2015).
20. Ipanema is a fashionable seashore neighborhood located in the South Zone of Rio de Janeiro.

2 Historicizing Ginga

1. Fado is a Portuguese musical genre that is strongly influenced by African, and possibly Afro-Brazilian, cultural heritages.
2. The historical kingdom of Matamba was located on the Cuango River northeast of Luanda, Angola.
3. Alencastro (2000) proposes that the religious conversion of Queen Ginga to Catholicism and her political defeat are celebrated as a triumph of the early Portuguese mercantilist (Roman Catholic) expedition.
4. *La Maravigliosa Conversione alla Santa Fede di Cristo della Regina Singa e del svo Regno di Matamba nell'Africa Meridionale* was published in Naples in 1669.
5. According to Thornton (1991, p. 38), her decision to "become a man" was a political strategy to overcome the illegitimacy of her gender.
6. *Istorica descrizione de' tre' regni Congo, Matamba et Angola* was written between 1654 and 1677 and was later published in Bologna in 1687.
7. See Alencastro (2000, pp. 280–1).

8. Even if one assumes that the Bantu verb "jinga," generally meaning to twist, turn, or oscillate, is the root of the Portuguese verb "gingar" and the nouns "ginga," "gingado," and "gingador," it might be equally impossible to determine exactly when, or in what context(s), these interconnected terms gain currency as an undulating or swayed way of moving characteristic of non-white (enslaved) bodies living in colonial Brazil.
9. The term "batuque" is also found in Portuguese descriptions of dance and music practices in Central Africa. In *Os Sertões da d´Africa* (1880), for example, Alfredo Sarnento employs batuque to reference endeavors found in the regions of Kongo, Ambriz, and Luanda, characterized by "a soft spoken wobbling of the body, paced by a small movement of the feet, the head, and arms" (quoted in Câmara Cascudo, 2001, p. 135).
10. See Foucault (1988).
11. As Dias (2001), Monteiro (2002), and Reis (2005) attest, the distinction between "honest" and "dishonest" activities were often evoked, but the exact position where one should draw the line dividing these two collections in colonial Brazil has been subject to interpretation. Since the sixteenth century, Catholic Brotherhood/Sisterhood organizations in Brazil (*Irmandades*) were in charge of administrating a wide range of celebrations organized around the Catholic liturgical calendar, when the enslaved population was granted a day off and exceptional rights. Their non-European dancing acts were subject to an array of reactions, despite the context in which they were enacted (e.g. orthodox Catholic, proto-Catholic, or "pagan" events).
12. As scholars in the field of both dance studies and post-colonial studies in Latin America seem to agree, this modern/colonial division of labor, knowledges, and subjectivity based on a (modern/colonial) idea of race, which gained visibility and validity during the Enlightenment, was equally informed by the previously existing division of labor according to gender and the canonic understanding of the rational mind in opposition to the objectified body.
13. *Die Baducca, in S. Paulo.* Lithograph by Johann Baptist von Spix, 1817–20. Image originally published in *Reise in Brasilien in den jahren 1817–1820* (Munich, 1823–31); Atlas (originally published 1831), Plate 27, upper right. This image portrays African/Afro-Brazilian enslaved men and women dancing under the supervision of a policeman. As I see it, their sexuality is depicted as desiring, but not honorable. Instruments depicted in this image include a large rasp (*reco-reco,* far left) and a balafon/marimba-type instrument.
14. *Batucca.* Lithograph by Louis-Jules-Frédéric Villeneuve after the original by Johann Moritz Rugendas, 1835. Originally published in *Voyage Pittoresque dans le Bresil* (Paris, 1835; also published in the same year in German). The illustration depicts a dance circle within a bucolic rural location, removed from urban and "civilized" spheres.
15. As the historian João José Reis proposes, the Bantus, Fon, Yorubas, etc. brought, along with their exploited physical energy, "the religious, aesthetic and moral values to create culture" (Reis, 2005, p. 201). It should be noted that this decentralized *recuperation-cum-invention* is not unique to Brazil, but is symptomatic of many former European colonies, whose subjugated populations underwent complex and multi-lateral processes of colonial violence under unique but ever-changing circumstances.

16. These ambulatory spectacles were enacted in exchange for donations within public areas (e.g. in front of churches or on the streets). According to J.J. Reis (1996), the primary goal of the public performances that the black-only Brotherhoods and Sisterhoods sponsored annually was to increase both the revenues of the religious organizations and the good feelings of their devoted practitioners. Among others, these donations were converted into funerary celebrations, ensuring that the deceased were buried with dignity. Socially, these festive events gave participants a sense of community.
17. Other proto-Catholic pageant-parades performed in Brazil include *Dança de São Gonsalo,* a public dance celebration in homage to the Portuguese St Gonçalo; *Pastoril Religioso* and *Pastoril Profano,* centered on the figure of shepherdesses from the Nativity Scene, enacted in a "religious" or "profane" manner; and *Folia do Divino,* centered on the figure of the Holy Spirit. Together, these authority-pleasing or donation-seeking narratives performed by converted Amerindians and Africans (and their offspring) have functioned as teaching and recruiting tools to their affiliated institutions.
18. Alencastro concludes that: "In philosophy, in literature, in popular religion, in reggae (songs), on the internet, in the feminist and black movements, the persistence of the pagan Ginga, the warrior Ginga against the religion of Whites, transfigure itself in the continuity of pre-Christian African culture, in the ogre of the Jesuit and capuchin preaching in Angola, and, more profoundly, of the five centuries of spectacular failure of missionary activity in Black West Africa" (2000, p. 282).
19. In his research, J.J. Reis cites a statute from one of those organizations, the Negro Minas Marri Congregation (*Congregação dos Pretos Minas Marri*) from 1786, where folias are clearly identified as a way of cheering up its current members as well as attracting potential new members and/or financial patrons (Reis, 1996, pp. 9–10).
20. According to Mario de Andrade, these city parades (*cortejos*) were constituted of "moveable pieces" such as songs that could be included or omitted, changed or not, based on the events of the day and the inspiration of the parade master (*mestre do cortejo*) (1982, p. 56).
21. Within rituals, worship is celebration (*siré*), a serious play moved by expressive arts (music, dance and costume). In Brazil, *Xirê* (possibly after the Yoruba *siré*) are celebratory events, which happen in the *aye* (also spelled *aiyê*), or material world, and constitute the meeting occasion for hot and cold principles, the material and the immaterial worlds: a terrain of continuous transformation and dynamic contestation.
22. Livio Sansone proposes that across Brazil and especially in black cities such as Salvador, Afro-Brazilians employ body language in addition to phenotype markers, clothing choices, and bodily modifications (e.g. hairstyles) to express their ethno-cultural identities (1999, p. 27).
23. See Soares (1999, 2004); Assunção (2005); and Talmon-Chvaicer (2008).
24. See Foucault (1995b).
25. According to Schwarcz (1993), at the end of the nineteenth century, the educated elites in Brazil prescribed two major solutions to the problem of how to remediate the social/moral "degeneration" of Brazilian society due to slavery/miscegenation. On the one hand, there were those who prescribed the racial "hygienization" of urban centers as a way of eradicating this

"illness," while on the other hand, there were those who supported the whitening of the society, via miscegenation, as a way of curing "sick bodies."

26. Radhika Mohanram states that: "In the rule of the superego and through the works of Plato through Locke, those humans who are white, male and normatively heterosexual prevail over those who are not" (1999, p. 50).
27. These personifications include, but are not limited to: (a) historical warriors associated with the political movement of resistance against the raced division of labor in Brazil such as Zumbi, the King of Palmares; (b) war deities that populate the pantheon of Afro-Brazilian religions (e.g. Candomblé), such as Xango, Ogun, and Oxossi; (c) historical figures such as Mestre Bimba and Mestre Pastinha; and (d) legendary capoeira characters with magical powers, such as Bezouro Mangangá.
28. As a cultural practice of resistance, Sansone concludes that: "It was developed as an anti-white institution and it is only in the last decades that the presence of whites in capoeira has become significant. The capoeira code of behavior, probably more than any other traditional institution, brings pride and self-esteem for the Brazilian blacks. An important theme of capoeira is the capacity of the Brazilian negro to resist and to fight against the slave holders" (1999, p. 32).
29. Marshal Florano Peixoto (President of Brazil, 1891–4) and José Maria da Silva Paranhos, Jr., Baron of Rio Branco (Minister of Foreign Affairs, 1902–12), are amongst the public figures who were allegedly trained in the eccentric "arts of capoeiragem" in their youth.
30. The first volume defending capoeira as a national gymnastic, *Guia da Capoeira ou Ginastica Nacional* (1907), is said to have been written by a naval officer who preferred to remain unidentified.
31. Author's translation. Original in Portuguese: "[j]oga-se os braços e o corpo em todos os sentidos em ginga, de modo a perturbar a attenção do adversário e preparar melhor o golpe decisivo" (Burlamaqui, 1928).
32. Given the enduring stigma against capoeira, Bimba opted to name his school the "Center of Regional Physical Culture."
33. Mestre Bimba's class of medical students was amongst the first to write extensively about capoeira and therefore has contributed to the archival of the practice as we know it today.
34. Mestre Bimba's teaching methodology, namely "Bimba's sequences," consisted of eight series of attack and defense movement phrases, to be first learned and rehearsed during training practice, and later employed within the improvised circle. For more information, see Rego (1968).

3 Understanding the Presence of Ginga in Samba Circles

1. Fanon, 1963, p. 57.
2. Rede Globo is one of the largest commercial TV networks in the world and is the leading producer of soap operas in the country. Nearly every household in Brazil is tuned to this station.
3. A recent study in Brazil (May–June 2013) conducted by the Instituto de Pesquisa Econômica Aplicada suggests that 65 percent of men in Brazil believe that women who wear revealing clothing deserve to be sexually attacked and/or raped (*Estado de São Paulo*, March 27, 2014).

4. Examples include *samba-chula*, characterized by call-and-response verbal improvisation and connected to samba circles; *samba-enredo*, a fast-paced samba particular of carnival parades or linear samba; *samba-canção*, a slow pace and sentimentalist samba most suitable to be danced in pairs; and the self-explanatory *samba-reggae*, *samba-rock*, and *samba-funk*.
5. As previously addressed in Chapter 1, this embodied logic of organization and production is informed by a set of Africanist aesthetic principles, identified in this book as polycentrism/polyrhythm, call-and-response, high-affect juxtaposition, serpentine pathways, apartness, coolness, serious play, and dissonance.
6. A machete is a small acoustic guitar originated in Portugal, similar to the ukulele. At the time that Machado de Assis wrote this chronicle, this guitar was associated with Afro-Brazilian rhythms and cultural practices, in opposition to the piano, a cultural symbol associated with European erudition. For the European-aspiring elites of Brazil, music produced with machetes as well as African drums and Moorish tambourines was perceived as inferior and distasteful.
7. Original text in Portuguese: "Todo ele acompanhava a gradação e variações das notas; inclinava-se sobre o instrumento, retesava o corpo, pendia a cabeça ora a um lado, ora a outro, alçava a perna, sorria, derretia os olhos ou fechava-os nos lugares que lhe pareciam patéticos. Ouvi-lo tocar era o menos; vê-lo era o mais. Quem somente o ouvisse não poderia compreendê-lo."
8. Dona Dalva named her samba group after the German tobacco factory of Suerdieck, where she worked and organized her first samba events in the late 1950s.
9. Caetano Veloso and Gilberto Gil, "Desde que o Samba é Samba," *Tropicalia 2*.
10. For Antonio Herculano Lopes, Rio de Janeiro had operated as a resonating box for an imagined "national culture" (2000, p. 2).
11. Maxixe, also known as West Indian burr gherkins (*Cucumis anguria*), is a vegetable that originated in West Africa and is abundant in Brazil. It is a staple of Afro-Brazilian cookery.
12. Briefly, *chula* may be understood as a form of spoken word battle (*desafio*) particularly popular in rural areas and performed in the context of circular dances, where two (or more) participants "compete" with one other by improvising call-and-response verses.
13. *See* also Curt Sachs (1937).
14. Chasteen also comments that, contrary to samba circles, at the time described in newspapers as a "traditional" and "wholesome" style of dancing, associated with the (colonial) past, the "infernal, unrestrained" proximity of maxixe dancers, as well as their "furious" and "incorrigible" sexuality, was often dismissed as a modern/urban social disease practiced by promiscuous people (2004, p. 45).
15. See Callois and Barash (2001).

4 Investigating the Articulation of Ginga in Capoeira Angola

1. Their bare feet symbolically indicates their enslaved condition, possibly as urban slaves for hire (*escravos de ganho*). Besides, the centralized depiction

of a trading exchange in the background further qualifies the environment where this "mock-combat" takes place: a territorialized black space, or "corner" (*canto*), controlled by a female food vendor (*quitandeira* or *baiana*).

2. The original German and the English translation available at: www.capoeira-palmares.fr/histor/maler_en.htm (date accessed April 23, 2015).
3. See Assunção (2005).
4. Here I follow the German dance scholar Rudolf von Laban's (1879–1958) theoretical use of the term "effort," first coined in his book *Effort* (1947) and further developed in collaboration with other artists and scholars. For Laban, effort is composed of two variables: one that relates to the physical description of movement, and is thus objectively measurable, and another that relates to psychosomatic experiences or the human faculties of perceiving reality, and is thus subjective but classifiable.
5. The basic structure of my movement analysis was developed over a course of five years (2005–10), while attending a wide range of capoeira classes, workshops, lectures, demonstrations, and performance events hosted at distinct schools and cultural centers in Brazil and in the USA. During this process, I also viewed and analyzed a variety of audio-visual materials (illustrations, photographs, LPs, CDs, videotapes, DVDs, and online videos) recorded in different parts of the world. For their contributions to my in-depth understanding of capoeira angola, I would like to acknowledge, in particular, the following learning institutions/master teachers: the International Foundation of Capoeira Angola (ICAF/FICA), led by Mestres Cobra Mansa (Cinézio Feliciano Peçanha), Valmir Damasceno, Jurandir Nascimento, and Silvio dos Reis; the Instituto Nzinga de Capoeira Angola, led by Mestres Janja (Rosângela Costa Araújo), Paulinha (Paula Barreto), and Poloca; ACANNE, led by Mestre Rene Bittencourt; Filhos de Angola, led by Mestre Laercio; and Semente de Angola, led by Mestre Jogo de Dentro.
6. In 1955, the center was transferred to Pelourinho, where they practiced until 1972, when the Instituto do Patrimonio Artistico e Cultural da Bahia evicted his school from its location.
7. Capoeira is one of the few martial arts that requires musical accompaniment. As such, the role of music in capoeira as well as the power relations exercised between players and musicians during a capoeira event are vital to the understanding of this multi-faceted practice as a whole. However, such discussions are beyond the scope of the discussion here.
8. In capoeira angola, the percussive orchestra is generally composed of three harmonized *berimbaus* (one-string bows), two *pandeiros* (tambourine), a *reco-reco* (guiro), an *agogô* (double bell), and an *atabaque* (drum).
9. The Brazilian "berimbau de barriga," or simply berimbau, is a gourd-resonated, braced musical bow of African origin. During colonial times, these instruments were often employed by street vendors as a way of drawing the attention of potential customers.
10. Most capoeira angola circles employ three berimbaus, distinguishable by their size and, subsequently, sound. They are: (a) *gunga*: larger gourd and thinner bow; (b) *viola*: smaller gourd and thicker bow; and (c) *medio*: medium-sized gourd and bow. Overall, the three berimbaus play call-and-response dialogues with one another.

11. For Downey, for example, the "Berimbau rhythm shapes the kinesthetic development of the game" (2005, p. 94).
12. Text originally recorded on Pastinha's album *Praticando Capoeira*, track 4, 3:27.
13. In practical terms, this "equivalence" depends on the level of skill of each player.
14. Here *malícia* should be understood beyond its literal translation, i.e. malice. In the context of capoeira, it often means a particular kind of intuitive cunningness, street-learned astuteness, playful trickery, or serious play (*ere* in Yoruba).
15. The legend goes that capoeira players used to hide a small blade in their hats, shoes, or handkerchiefs in order to literally "mark" their opponent's face or upper body with a (superficial) cut.
16. These metaphors are also widely available within capoeira's oral tradition.
17. Although the implicit "fight," "dance," or "play" qualities are commonly understood in practice, they are seldom verbalized in these discrete categories. Among capoeira players, maneuvers are referred to by their contextual application (e.g. "to mock him" or "to scare him") or their codified names ("hammer" or "stingray's tail"), which vary greatly from school to school and from region to region.
18. The harmonious combination of seductive dance, playful mockery, and aggressive fight that appear and disappears, in circular motions, may be employed to balance out the fixed otherness attributed to (or imagined of) non-white bodies.
19. In capoeira, the expression "at the foot of the berimbau" (*ao pé do berimbau*) indicates the space next to the feet of the person playing that instrument during a roda de capoeira. The "foot of the berimbau" is regarded as the "entrance/exit" of the playing field and thus the "starting/ending point" of capoeira games.
20. In Brazil, the term *negaça* has multiple and correlated meanings: enticement, allurement; bait, decoy; provocation; feint; take-in, trap; mistake. Within the realm of capoeira, more specifically, it refers to the act of negating or dissimulating intent through bodily action.

5 What is it about the Baiana?

1. "O Que é Que a Baiana Tem?" was especially composed by Doryval Caymmi for the musical film *Banana da Terra* (1939).
2. The word "baiana" literally means a woman from the state of Bahia, where an estimated 80 percent of the population is black. Salvador, Bahia's capital, has the largest percentage of African descendants outside Africa. It has thus been nicknamed the "Black Rome."
3. The expression"pano da costa" literally means fabric from the (West African) coast. In Brazil, it references a long rectangular shawl that Afro-Brazilian women wore over their shoulders, originally identifying their ethnicity and/or social status.
4. Neto was a Marxist who fought for the impendence of Angola. When that nation became independent from Portugal in 1975, the poet and activist was proclaimed the first President of the People's Republic of Angola.

5. The poem is available at: http://www.agostinhoneto.org/index.php?option=com_content&view=article&id=603:quitandeira&catid=45:sagrada-esperanca&Itemid=233 (date accessed April 28, 2015).
6. Escravos de ganho were allowed to collect the profits of their labor after paying a fixed daily/weekly rate to their owners. Their earnings could be used to purchase anything from food and clothing to their own freedom.
7. According to the poet and folklorist Cecilia Meirelles (1983), when going to work, these street vendors would balance a large wooden tray over their heads, where they laid their merchandise, as well as a small stove, a glass lantern, utensils, napkins, etc. When shopping, the tray was replaced by a basket, in which they carried anything from groceries and spices to their own shoes.
8. Acarajé is a rich dish made with white bean cake deep-fried in palm oil (*dendê*) and stuffed with shrimp, salad, ochre, and a fish paste (*vatapá*).
9. The nickname was given by Heitor dos Prazeres, a samba composer and visual artist who lived at Cidade Nova. It illustrates the progressive gentrification of Rio de Janeiro during the second half of the nineteenth century.
10. Tia Ciata was the grandmother of the samba composer Bucy Moreira (1909–82). Tia Amelia do Aragão, Amélia Silvana de Araujo, was the mother of the violinist and samba composer Donga (Ernensto Joaquim Maria dos Santos, 1890–1974). Perciliana Maria Constançal, or Tia Perciliana, was the mother of the famous samba composer, singer, and dancer João da Baiana (1887–1974).
11. For an in-depth discussion regarding the "Bahian colony" in Rio de Janeiro, see Moura (1983); Tinhorao (1988b); N. Lopes (1992); Cabral (1996); Sodré (1998); and Velloso (1990).
12. See J.J. Reis (2005).
13. In the version recorded by Odeon in 1902, the lyrics were edited. Further disconnecting that racially ambiguous character from the old black vendor, or "real" Sabina, the updated lyrics state that "this way they proved how much they like the *mulata*" ("deste modo provaram como gostam da mulata," quoted in Gomes and Seigel, 2002, p. 179, author's translation).
14. The album *Isso É Bom* (*This is Good,* 1902), interpreted by Bahiano (Manuel Pedro dos Santos), was recorded at Frederico Figner's Casa Edison and was released by his record company Odeon.
15. Chasteen notes that while the idealized "brunette" (*morena*) of Rio's carnaval was in the 1850s a "rare and ratified literary image cultivated in white clubs," by the turn of the twentieth century, "thousands of flesh-and-blood black women literally danced down the middle of the street while these songs were sung" (1996, p.43).
16. Martha Rocha, a German Brazilian born in Bahia, was elected Miss Brazil in 1954. That same year, Rocha represented Brazil in the Miss Universe contest in Long Beach, California, wearing a gown inspired by the "typical" baiana attire.
17. This article was also published in the newspaper *Estado de São Paulo* that same year, under the title "Carmen Miranda dada."

6 Brazilian Bodies and Nationalism in Dance

1. Although the first re-staging of *Swan Lake* in Brazil would only appear in 1959, the European-aspiring elites also considered this ballet a quintessential example of its genre.

2. Wisnik remained very present throughout the choreographic montage of *Nazareth* and has, since then, provided Grupo Corpo's audience with his insightful interpretations of the project. His discussion regarding the relationship between the work of Machado de Assis and Ernesto Nazareth was published in a seminal essay "Machado Maxixe: O Caso Pestana" (Wisnik, 2004).
3. Exceptions include *Sete ou Oito Peças para um Ballet* (1994), composed by Philip Glass and recorded by Uakti, and *Lecuona* (2004), choreographed to the music of the Cuban composer Ernesto Lecuona.
4. Here I employ the Portuguese term "balé" to reference the various overlapping subcategories of the ballet technique, which have merged and interacted in Brazil.
5. In Brazil, companies historically engaged in such discussions include the historical *Teatro Experimental do Negro*, founded in 1944 by Abdias do Nascimento, and, more recently, *SeráQ* and *Rui Moreira Companias de Dança*, both directed by Rui Moreira (a former Grupo Corpo dancer).
6. Other examples include Ariano Suassuna's *Balé Armorial do Nordeste*, founded in 1976, and *Balé Folclórico da Bahia*, founded in 1988.
7. The Brazilian-Portuguese verb "encaixar" literally means to set or fit something inside a box or into a groove. Figuratively, however, it connotes a perfect match between discreet elements that falls into place, as in the English expression "hand in glove." In dance, encaixar references the turn-out rotation of the hips characteristic of classical ballet, which "locks in" the pelvis.
8. The Italian ballet master Carlos Blasis proposes in his *Code of Terpsichore* that, "the head, arms, the hands, legs, feet, in short all parts of the body are rendered symmetrical, pliant and graceful" (1976 [1831], pp. 25–6).
9. The Brazilian-Portuguese verb "desencaixar" literally means "to take something out of a box or out of line, to dislocate or disjoin parts." As a metaphor, desencaixar connotes something out of place, a fragment or dissonance. In dance, the verb is used in reference to a loose or disjointed movement of the pelvis, sideways or backwards, which from a ballet perspective deflects the upright alignment of the body and fragments the unity of the core.
10. Volúsia was a ballet-trained dancer best known in Brazil as the creator of the "bailado brasileiro," which featured auto-exoticized representations of local dance practices and character types, most of which were centered on Afro-Brazilian traditions. In the 1930s and 1940s, the Ministry of Health and Education hired Volusia to teach "folk" dance classes and choreograph dance recitals. Following the footsteps of Miranda, Volúsia appeared in Abbott and Costello's comedy *Rio Rita* (1942) and, while shooting that film, was featured on the cover of *Life Magazine* (September 22, 1941).
11. The work of Rudolf von Laban, Mary Wigman, and other subscribers of the German expressionist dance (*Ausdruckstanz*) is the primary influence of this subgenre of modern dance in Brazil.
12. The work of George Balanchine (1904–83) has influenced this subgenre of *balé* in Brazil.
13. Mercedes Baptista was the first Afro-Brazilian dancer to take ballet classes at the Municipal Theater "Escola de Bailados." Not too far from the exoticism available in the 1930s "Brazilian ballets," the *Ballet Folclórico Mercedes Baptista* toured internationally in the late 1950s with stage adaptations of

a variety of dancing practices from Afro-Brazilian religious ceremonies to samba.

14. See also Daniela Reis (2005a).

7 What is it about Grupo Corpo?

1. For Wisnik, the piano functioned symbolically in colonial Brazil as an aristocratic piece of furniture that, along with the mass-produced music sheets, placed salon music playing and dancing as central activities within middle-class secluded living rooms. For the author, the piano brings, in and of itself, a prestigious fragment of Europe into the house of European-aspiring elites in Brazil. It functions as a mixture of a metonym of modern civilization and the ornament of the private realm of an enslaved society, where it entertain the "respectable ladies" confined to their domestic settings (Wisnik, 2004, p. 54).
2. For Wisnik, Ernesto Nazareth, Machado de Assis, Rodrigo Pederneira, and himself are artists whose work may be situated between "the popular" and "the erudite." Their artistic creations are invested in the blurred space they construct between these two imagined ideals.
3. The Polka ("Cross, Peril!"); The Waltz ("Concerto Impromptu"); Flora's Sonata; The Maxixe ("What is this?"); Baroque Sonata ("Sagaz"); The Maxixe ("Golden Blue"); Waltz of Mirrors; Chorinho ("Dandy"); Funeral March/ Flora's Sonata; and Tool ("Tool").
4. For Wisnik, this middle ground associated with the concept of "mulatto-ness" or "gray-ness" remains within Brazilian society as a type of "hidden secret." It is "debated in very deep levels, because within it lies the knot that ties up the formally impermeable terms of the social structure – the slave and the slave master – across the proliferating chain [*elo proliferante*], that which is obvious and that which is hidden, between slavery and sexuality, which 'invents' socially and culturally, in Brazil, the mulatto. This knot, I should say, is pure ambivalence, because – beyond the [male] slave master and the enslaved female and beyond the free white man – the mulatto man, at the edge of this process, is [located] at the borderland of [both] exclusion and inclusion [The mulatto is] the piece, neither rejected nor accepted, which holds the inconfessable secret of the whole. This position is homologous, in its own term, to the place occupied by Africanist popular music, between rejection and seduction, irrepressible indexes of Brazilian life, which will become later on the celebrated icons of Modern Brazil, and the privileged path of its symbolization" (Wisnik, 2004, p. 64, author's translation).
5. In *Dança no Brasil*, Ida Vicenzia summarizes that "the ballet's singularity and its lack of preoccupation with content makes it the least provocative of Pederneiras's successful career" (1997, p. 121).
6. The *Biennial of Dance* took place from September 3 to 29, 1994. It featured a wide range of companies, from Alvin Ailey to Bill T. Jones.
7. My analysis is based on both my personal memory of *Breu*'s rehearsals and staged performances, both in 2008 (Brazil) and in 2009 (US), as well as excerpts of the ballet's archived performance (2008, DVD format) featuring its original cast.
8. Lenine is a Grammy Award-winning artist from Recife, Pernambuco.

9. *Cabloquinho*, from *cabloco* (a mixture of Amerindian and European ethnicities), is a rhythm inspired by folk ceremonies specific to the state of Pernambuco.
10. Colonized by Portugal and later invaded by Dutch settlers, Recife was one of the major ports in Brazil during colonial times. Located in the northeast of Brazil, a sugar cane plantation region, Recife's port received the enslaved labor force and disbursed them across the country.
11. After a series of transformations, today frevo looks like an acrobatic dance performed with tiny umbrellas, where men dress and act like jesters and women wear short and sexy flapping skirts.
12. In this scenario, the dancers' conflicting interactions give the impression of Ptolemy's purgatory, where the same action is repeatedly endlessly.
13. This is not entirely an innovation. Horizontality was already present in *Oncotô* (2005).
14. Alejandro Ahmed, the choreographer of Grupo Cena 11, has engaged his company in an unorthodox movement research through which they developed an extremely violent dance technique called "physical perception." According to Greiner, this technique "is based on the attempt to control the most out-of-control situations, like a violent fall or a crash. The exercises he derived with his company strengthened Ahmed's bones while eliminating the risk of injury. However, the risk was not completely eliminated: through the audience's eyes, the movements still seem very risky and dangerous" (2007, p. 142).
15. Joseph Roach presented this rhetorical question during a public lecture at UCLA, May 2008.
16. Randy Martin poses that "while no single work of art can effect historical transformation, it can provide the occasion to reflect on such possibilities" (1998, p. 184).

Conclusion: The Pride and Shame of Being a Swing Nation

1. The "Kilombo" (an encampment in Kimbundu) is named after the historical maroon communities (e.g. *Quilombo dos Palmares* in Pernambuco) found by freed and escaped enslaved Africans during the colonial era in Brazil.
2. Mestre Cobra Mansa, or Cinézio Feliciano Peçanha, is one of the founders of the International Capoeira Angola Foundation (ICAF; FICA in Portuguese) and perhaps one of the most widely known and respected practitioners of this Afro-Brazilian martial art. In the 1980s, Mestre Cobra Mansa moved to Salvador and, along with Mestre Moraes, started a groundbreaking organization, the already-mentioned Grupo Capoeira Angola Pelorinho (GCAP).
3. Originally known as Saint Anthony Fortress, this colonial building is a kind of "Mecca" for practitioners of capoeira worldwide.
4. This march paid homage to *Azoany*, an Afro-Brazilian deity from the Jeje tradition associated with physical health and the prevention of skin maladies and contagious diseases, including AIDS. During the colonial era in Brazil, Azoany, also known as *Obaluaê* (Yoruba tradition), was syncretized with the Catholic St Lazarus and, since then, Brazilians from both religious affiliations honor these entities on August 16. In Salvador, the final destination of the Azoany March is the St Lazarus feast, a syncretic celebration that

includes a Catholic mass and a Candomblé blessing ceremony augmented by a purifying "shower" of popcorn, followed by eating, drinking, and samba dancing.

5. Available at: http://blogs.cultura.gov.br/blogdarouanet/category/publicacoes (date accessed May 7, 2015).
6. Available at: www.atitudebrasil.com/site/wp-content/uploads/2011/10/rouanet_web.pdf (date accessed May 7, 2015).
7. Exemples include the Popular Centers of Culture of the Students National Union (*Centros Populares de Cultura da União Nacional dos Estudantes*) and the Popular Culture Movement (*Movimento de Cultura Popular*; AP 2011).
8. These include performing arts; classical or instrumental music; major visual art exhibition; artistic, literary, or humanistic books; donations of collections to public libraries as well as staff training and acquisition of equipment for the maintenance of their facilities; museums; film collections; the preservation and dissemination of audiovisual collections; and the preservation of tangible and intangible cultural heritages (AP 2011). The film industry is supported by a similar, although separate, legislation. Similar pieces of legislations are also implemented at the state and municipal levels.
9. See www.wagnerschwartz.com/transremontagem.html (date accessed May 7, 2015).
10. Discussion originally posted in 2012 at: www.nucleododirceu.com.br/colecao-brasilidade-apresenta-o-gesto (date accessed February 17, 2013).
11. Full text in Portuguese: "O discurso multicultural também é um outro extremo, o blasê de 'o brasil é o mundo minha gente.' O clichê da mistura. Nas Olimpiadas vai ter africa e japão no shape carnaval-arraso. Essa é a nossa identidade brasileira, essa idéia de que somos muitos ao ponto de poder ser 'quase qualquer um qualquer coisa. Viva o Brasil multietnico, miscigenado, plural!!' Sabia não? Os taxistas da Glória já estão fazendo campanha pra Debora Colker dirigir a abertura dos jogos. E agora que estamos sendo convocados a mostrar o que entendemos por brasilidade ao mundo que corpo é esse? que dança é essa?" Text originally posted in 2012 at: www.nucleododirceu.com.br/colecao-brasilidade-apresenta-o-gesto (date accessed February 17, 2013).
12. Full text in Portuguese: "A partir de um roteiro que nos foi encaminhado, desenhamos todos os figurinos. Procuramos valorizar manifestações populares através da gambiarra como recurso de linguagem artística. A Gambiarra ou improvisação é um dos traços que caracterizam a produção artística no Brasil, com a assimilação pela arte de procedimentos e estéticas das ruas e do povo ... O Brasil contemporâneo se fez pulsante por meio do conceito da antropofagia, do modernismo e, posteriormente, da Tropicália. A antropofagia na arte brasileira é a percepção das influências estrangeiras, sua compreensão e deglutição em uma apropriação que mantém apenas o que interessa à criação local, de modo consciente e refletido. Se a apropriação de elementos de outras culturas e países é aceitável por esta elite cultural que critica a cerimônia, o que nos impede de sermos antropófagos dos nossos próprios clichês e valores?" Text originally published on Jum Nakao's Facebook page on August 17, 2012: https://www.facebook.com/jumnakao?fref=ts (date accessed February 17, 2013).

13. At this point, this future project seeks to measure and qualify the extent to which this interaction may: "Permit individuals to effect by their own means or with the help of others a certain number of operations on their own bodies and souls, thoughts, conduct, and way of being, so as to transform themselves in order to attain a certain state of happiness, purity, wisdom, perfection, or immortality" (Foucault, 1988, p. 19).

Bibliography

Publications

Abdalla, Antônio Carlos (ed.) (2011) *Lei Rouanet Percursos e Relatos*. Atitude Brasil. Available at: www.atitudebrasil.com/site/wp-content/uploads/2011/10/rouanet_web.pdf. (date accessed May 11, 2015).

Abib, Pedro (2004) *Capoeira Angola. Cultura popular e o jogo dos saberes na roda*. Salvador: UFBA.

Abreu, Frederico José de (2003) *O barracão do Mestre Waldemar*. Salvador: Zarabatana.

Abreu, Martha (1994) *Festas religiosas no Rio de Janeiro. Perspectivas de controle e tolerância no Século XIX*. Rio de Janeiro: Estudos Históricos.

Agamben, G. (2005) *State of Exception*. University of Chicago Press.

Albright, Ann Cooper (1997) *Choreographing Difference: The Body and Identity in Contemporary Dance*. Hanover, NH: Wesleyan University Press.

Albright, Ann Cooper and Gere, David (eds) (2003) *Taken by Surprise: A Dance Improvisation Reader*. Middletown, CT: Wesleyan University Press.

Alencastro, Luiz Felipe de (2000) *O tratado dos viventes, formação do Brasil no atlântico sul, séculos XVI e XVII*. São Paulo: Companhia das Letras.

Amadei, Yolanda (2006) "Correntes migratórias da dança: modernidade brasileira." In Mommensohn, Maria and Petrella, Paulo (ed.), *Reflexões sobre Laban, o Mestre do Movimento*. São Paulo: Summus.

Amaral, Rita (2002) *Xire! O modo de crer e de viver no Candomblé*. Rio de Janeiro: Pallas.

Anderson, Benedict (1991) *Imagined Communities: Reflections on the Origin and Spread of Nationalism*. London and New York: Verso.

Andrade, Mario de (1928) *Macunaíma, o herói sem nenhum caráter*. São Paulo: Eugenio Cupolo.

——. (1972) *Ensaio sobre a música brasileira*. São Paulo: Martins.

——. (1982) *Danças dramáticas do Brasil*, 2 vols, vol. 1. Belo Horizonte: Editora Itatiaia.

Antônio, João (1991) *Zicartola e que tudo mais vá pro inferno!* São Paulo: Editora Scipione.

Araújo, Rita de Cássia Barbosa de (1996) *Festas: máscaras do tempo, entrudo, mascarada e frevo no carnaval do Recife*. Recife: Fundação de Cultura Cidade do Recife.

Assunção, Matthias Röhrig (2005) *Capoeira: The History of an Afro-Brazilian Martial Art*. New York: Taylor & Francis.

Austin, J.L. (1962). *How to Do Things with Words: The William James Lectures Delivered at Harvard University in 1955*. Oxford University Press.

Badinter, Elisabeth (1997). *XY: On Masculine Identity*. New York: Columbia University Press.

Bakhtin, Mikhail (1984) *Rabelais and His World*, Helene Iswolsky (trans.). Bloomington: Indiana University Press.

Banes, Sally (1998a) *Dancing Women: Female Bodies on Stage*. New York: Routledge.
——. (1998b) *Subversive Expectations: Performance Art and Paratheater in New York, 1976–85*. Ann Arbor: University of Michigan Press.
Barbosa, Livia Neves de H. (1992) *O jeitinho brasileiro: a arte de ser mais igual que os outros*. Rio de Janeiro: Editora Campus.
Barbosa, Wallace de Deus (2007) *Inventário para a registro e salvaguarda da capoeira como patrimônio cultural do Brasil*. Brasilia: IPHAN.
Barnes, Sandra T. (1997) *Africa's Ogum: Old World and New African Systems of Thought*, 2nd edn. Bloomington: Indiana University Press.
Bastide, Roger (2005) *O candomblé da Bahia Rito nagô*. São Paulo: Companhia das Letras.
Bastos, Wilson de Lima (1977) *Danças no registro do folclore brasileiro*. Juiz de Fora: Centro de Estudos Sociológicos de Juiz de Fora.
Bataille, Georges (1986) *Erotism: Death and Sensuality*. San Francisco: City Lights Books.
Batenson, Gregory (1972) *Steps to Ecology of Mind*. New Jersey: Jason Aronson Inc.
Béhague, Gerard (1998) "Afro-Brazilian Traditions." In Dale A.Olsen and Daniel E. Sheely (eds), *The Garland Encyclopedia of World Music. 2: South America, Mexico, Central America, and the Caribbean*. New York: Garland.
Berenstein-Jacques, Paola (2001a) "The Aesthetics of the Favela: The Case of an Extreme." In K. Thornton (ed.), *Transforming Cities, Design in the Favelas of Rio de Janeiro*, London: AA Publications.
——. (2001b) *Estética da Favela. A arquitetura das favelas através da obra de Helio Oiticica*. Rio de Janeiro: Casa da Palavra.
Blasis, Carlo (1976 [1831]) *The Code of Terpsichore: A Practical and Historical Treatise, on the Ballet, Dancing, and Pantomime*. Brooklyn, NY: Dance Horizons.
Boal, Augusto (1985) *Theatre of the Oppressed*. New York: Theatre Communications Group.
Bocage, Manuel Maria Barbosa du. (1908) *O livro dos sonetos*. Lisbon, Rio de Janeiro: A Editora.
Bogéa, Inês (ed.) (2001). *Oito ou nove ensaios sobre o Grupo Corpo*. São Paulo: Instituto Tomie Ohtake, Cosac & Naify.
Browning, Barbara (1995) *Samba: Resistance in Motion*. Bloomington: Indiana University Press.
——. (1998) *Infectious Rhythm: Metaphors of Contagion and the Spread of African Culture*. New York: Routledge.
Bruhns, Heloisa (2000) *Futebol, carnaval e capoeira: entre as gingas do corpo brasileiro*. Campinas: Papirus Editora.
Burt, Ramsey (1998) *Alien Bodies: Representations of Modernity, "Race," and Nation in Early Modern Dance*. New York: Routledge.
——. (2006) *Judson Dance Theater: Performative Traces*. New York: Routledge.
Butler, Judith (2004) "Performative Acts and Gender Construction." In H. Bial (ed.), *The Performance Studies Reader*. New York: Routledge.
Cabral, Sérgio (1996) *As Escolas de Samba do Rio de Janeiro*. Rio de Janeiro: Lumiar.
——. (1974) *Samba Schools: What, Who, How, When and Why*. Rio de Janeiro: Fontana.
Caillois, R. and Barash, M. (2001) *Man, Play, and Games*. Urbana: University of Illinois Press.
Câmara Cascudo, Luis da (1972) *Dicionário do folclore brasileiro*. Rio de Janeiro: MEC/Instituto Nacional do Livro.

——. (1987) *História dos nossos gestos*. São Paulo: EdUSP.
——. (2001) *Made in África*. São Paulo: Global.
Caponi-Tabery, Gena (ed.) (1999) *Signifyin(g), Sanctifyin' & Slam Dunking: A Reader in African American Expressive Culture*. Amherst: University of Massachusetts Press.
Cardonega, Antonio de Oliveira de (1942) *História geral das guerras angolanas, 1681*, 3 vols. Cônego José Mathias Delgado (ed. vols 1 and 2) and Manuel Alves da Cunha (ed. vol. 3). Lisbon, 1972.
Carneiro, Edison (1961) *Samba de umbigada*. Rio de Janeiro: MEC.
——. (1975) *Capoeira*. Rio de Janeiro: Campanha de Defesa do Folclore Brasileiro.
——. (1982) *Folguedos tradicionais*. Rio de Janeiro: Funarte, INF.
Carvalho, José Jorge de Carvalho (1994) "Black Music of All Colors: The Construction of Black Ethnicity in Ritual and Popular Genres of Afro-Brazilian Music." In Gérard Béhague (ed.), *Music and Black Ethnicity. The Caribbean and South America*. New Brusnwick: Transaction Publishers.
Carybé (1962) *As sete portas da Bahia*. São Paulo: Martins.
Cassoli, Camilo, Falcão, Luiz Augusto and Aguiar, Rodrigo (2007) *Frevo, 100 anos de folia*, Henrik Carbonnier (trans.). São Paulo: Timbro.
Castro, Ruy (2003) *Carnaval no fogo: crônica de uma cidade excitante demais*. São Paulo: Companhia das Letras.
——. (2005) *Carmen: uma biografia*. São Paulo: Companhia das Letras.
Castro-Gómez, Santiago (2008) "(Post)Coloniality for Dummies: Latin American Perspectives on Modernity, Coloniality, and the Geopolitics of Knowledge." In M. Moraña, E. Dussel, and C. Jáuregui (eds), *Coloniality at Large: Latin America and the Postcolonial Debate*. Durham, NC: Duke University Press.
Cavazzi, Giovanni Antonio (1687) *Istorica descrizione de' tre' regni Congo, Matamba et Angola: sitvati nell' Etiopia inferiore occidentale e delle missioni apostoliche esercitateui da religiosi Capuccini*. Bologna: Giacomo Monti.
Caymmi, Stella (2001). *Dorival Caymmi*, o mar e o tempo. São Paulo: Editora 34.
Certeau, Michel de (1984) *The Practice of Everyday Life*. Berkeley: University of California Press.
Chasteen, John (2004) *National Rhythms, African Roots: The Deep History of Latin American Popular Dance*. Albuquerque: University of New Mexico Press.
Chaui, Marilena (2003) *Convite à Filosofia*. São Paulo: Ática.
Chaves, Rita, Secco, Carmen Lucia Tindo and Macedo, Tania (eds) (2006) *Brasil África: como se o mar fosse mentira*. São Paulo: Ed. Unesp.
Crook, Larry and Johnson, Randal (eds) (1999) *Black Brazil: Culture, Identity, and Social Mobilization*. Los Angeles: UCLA Latin American Center Publications.
Daly, Ann (1995) *Done into Dance: Isadora Duncan in America*. Bloomington: Indiana University Press.
DaMatta, Roberto (1987) *A casa e a rua: espaço, cidadania, mulher e morte no Brasil*. Rio de Janeiro: Guanabara.
——. (1997) *Carnavais, Malandros e Heróis, para uma sociologia do dilema brasileiro*. Rio de Janeiro: Rocco.
Daniel, Yvonne (2005) *Dancing Wisdom: Embodied Knowledge in Haitian Vodou, Cuban Yoruba, and Bahian Candomblé*. Urbana: University of Illinois Press.
Debret, Jean Baptiste (1954) *Viagem pitoresca e histórica ao Brasil*, 2 vols, Sérgio Milliet (trans.). São Paulo: Livraria Martins Editora.
Decânio Filho, Ângelo A (1996) *A Herança de Mestre Bimba*, 2nd edn. Salvador: self-published.

——. (1997) *A Herança de Pastinha*, 2nd edn. Salvador: Coleção São Salomão 3.

DeFrantz, Thomas F (2004) "The Black Beat Made Visible: Hip Hop Dance and Body Power." In Andre Lepecki (ed.), *Of the Presence of the Body: Essays on Dance and Performance Theory*. Middletown: Wesleyan University Press.

——. (2004) *Dancing Revelations: Alvin Ailey's Embodiment of African American Culture*. New York: Oxford University Press.

Delgado, Celeste Fraser and Muñoz, José Esteban (1997) *Everynight Life: Culture and Dance in Latin/o America*. Durham, NC: Duke University Press.

Desch Obi, T.J (2008) *Fighting for Honor: The History of African Martial Art Traditions in the Atlantic World*. Columbia, SC: University of South Carolina Press.

Desmond, Jane C (ed.) (1997) *Meaning in Motion: New Cultural Studies of Dance*. Durham, NC: Duke University Press.

——. (ed.) (2001) *Dancing Desires: Choreographing Sexualities on and off the Stage*. Madison: University of Wisconsin Press.

Diamond, Elin (1993) "Mimesis, Mimicry, and the 'True-Real'." In Lynda Hart and Peggy Phelan (eds), *Acting Out: Feminist Performances*. Ann Arbor: University of Michigan Press.

Dias, Luiz Sérgio (2001) *Quem tem medo da capoeira? Rio de Janeiro, 1890–1904*. Rio de Janeiro: Prefeitura da Cidade do Rio de Janeiro, Secretaria Municipal das Culturas, Departamento Geral de Documentação e Informação Cultural, Arquivo Geral da Cidade do Rio de Janeiro, Divisão de Pesquisa.

Douglas, Mary (1976) *Pureza e Perigo*. São Paulo: Perspectiva.

Downey, Greg (2005) *Learning Capoeira: Lessons in Cunning from an Afro-Brazilian Art*. Oxford University Press.

Drewal, H.J. (1999) "Art History, Agency and Identity: Yoruba Transcultural Currents in the Making of Black Brazil." In Larry Crook and Randal Johnson (eds), *Black Brazil: Culture, Identity, And Social Mobilization*. Los Angeles: UCLA Latin American Center Publications.

——. (1999) "Memory and Agency: Bantu and Yoruba Arts in Brazilian Culture." In N. Mirzoeff (ed.), *Diaspora and Visual Culture*. London: Routledge.

Drewal, H.J. and Mason, J. (1998) *Beads, Body, and Soul: Art and Light in the Yorùbá Universe*. Los Angeles: UCLA Fowler.

Drewal, H.J., and Pemberton III, J. with Abiodun, Rowland (eds) (1989) *Yoruba: Nine Centuries of African Art and Thought*. New York City: Center for African Art.

Drewal, H J., Pemberton III, J. and Abiodun, Rowland (2003) "The Yoruba World." In Wole Soyinka and Simon Gikandi (eds), *Death and The King's Horseman*. New York: W.W. Norton.

Drewal, Margaret Thompson (1992) *Yoruba Ritual: Performers, Play, Agency*. Bloomington: Indiana University Press.

——. (2003) "Improvisation as Participatory Performance: Egungun Masked Dancers in the Yoruba Tradition." In A.C. Albright and D. Gere (eds), *Taken by Surprise: A Dance Improvisation Reader*. Middletown: Wesleyan University Press.

Du Bois, William Edward Burghardt (1996) *The Souls of Black Folk*. New York: Penguin.

Duarte, Ruy (1968) *História social do frevo*. Rio de Janeiro: Leitura.

Dunn, Christopher (2001) *Brutality Garden: Tropicália and the Emergence of a Brazilian Counterculture*. Chapel Hill: University of North Carolina Press.

Durst, Rogério (1985) *Madame Satã, com o diabo no corpo*. São Paulo: Brasiliense.

Efegê, Jota (1974) *Maxixe, a dança excomungada*. Rio de Janeiro: Conquista.

Fanon, Fantz (1963) *The Wretched of the Earth*. New York: Grove Press.

——. (1967) *Black Skin, White Masks*, Charles Lamm Markmann (trans.). New York: Grove/Atalntic Inc.

Faro, Antônio José (1988) *A dança no Brasil e seus construtores*. Rio de Janeiro: FUNACEN.

Ferreira, Aurélio B. de Holanda (1986) *Novo dicionário da língua portuguesa*, 2nd edn. Rio de Janeiro: Nova Fronteira.

Flusser, Vilém (1998) *Fenomemologia do brasileiro: em busca de um novo homem*. Rio de Janeiro: EDUERJ.

——. (2014) *Gestures*. Minneapolis: University of Minnesota Press.

Foster, Susan Leigh (1986) *Reading Dancing: Bodies and Subjects in Contemporary American Dance*. Berkeley: University of California Press.

——. (1995) "An Introduction to Moving Bodies: Choreographing History." In Susan L. Foster (ed.), *Choreographing History*. Bloomington: Indiana University Press.

——. (2011) *Choreographing Empathy: Kinesthesia in Performance*. New York: Routledge.

——. (2011a). *Worlding Dance*. Basingstoke: Palgrave Macmillan.

——. (ed.) (1995) *Corporealities: Dancing, Knowledge, Culture, and Power*. New York: Routledge.

Foucault, Michel (1978) *History of Sexuality I: An Introduction*, Robert Hurley (trans.). New York: Vintage Books.

——. (1980) *Power/Knowledge: Selected Interviews and Other Writings: 1972–1977*. New York: Village Books.

——. (1986) *The Use of Pleasure. Volume 2 of the History of Sexuality*. New York: Vintage Books.

——. (1988) *Technologies of the Self: A Seminar with Michael Foucault*. Amherst: University of Massachusetts Press.

——. (1995a) *As Palavras e as Coisas*. São Paulo: Martins Fontes.

——. (1995b) *Discipline and Punish: The Birth of the Prison*, 2nd edn, Alan Sheridan (trans.). New York: Vintage Books.

Freyre, Gilberto (1933) *Casa Grande & Senzala*. Rio de Janeiro: Maia and Schmidt.

——. (1936) *Sobrados e Mucambos. Decadência do partiarcado rural no Brasil*. São Paulo: Compania Editora Nacional.

——. (1963) *The Mansions and the Shanties: The Making of Modern Brazil*, Harriet De Onís (trans. and ed.). New York: Knopf.

——. (1974) *The Gilberto Freyre Reader*. New York: Knopf.

Fry, Peter (1982) *Para inglês ver: identidade e política na cultura brasileira*. Rio de Janeiro: Zahar Editores.

Fuss, Diana (1995) *Identification Papers*. New York: Routledge.

Garcia, Cindy (2013) *Salsa Crossings: Dancing Latinidad in Los Angeles*. Durham, NC: Duke University Press.

Gates, Henry Louis (1988) *The Signifying Monkey: A Theory of African-American Literary Criticism*. New York: Oxford University Press.

Geertz, Clifford (1978) *A Interpretação das Culturas*. Rio de Janeiro: Zahar Editores.

Giacomini, Sônia (1988) *Mulher e escrava*. Rio de Janeiro: Vozes.

Giannotti, Marco (2001) "Reflexões sobre o corpo e o espaço." In Inês Bogéa and José Luiz Pederneiras (eds), *Oito ou nove ensaios sobre o Grupo Corpo*. São Paulo: Cosac & Naify.

Giffoni, Maria A (1964) *Danças folclóricas brasileiras*. Rio de Janeiro: Melhoramentos.

Gil, José (2005) *Movimento total, o corpo e a dança*, Miguel Serras Pereira (trans.). São Paulo: Iluminuras.

Gilbert, Helen and Tompkins, Joanne (1996) *Post-colonial Drama: Theory, Practice, and Politics*. New York: Routledge.

——. (1997) *Exer(or)cising Power: Black Bodies in the Black Public Sphere. Dance in the City*, H. Thomas (ed.). New York: St Martin's Press.

Gilroy, Paul (1993) *The Black Atlantic: Modernity and Double Consciousness*. London: Verso.

Góas Dantas, Beatriz (1988) *Vovó Nagô e Papai Branco. Uso e abuso da Africa no Brasil*. Rio de Janeiro: Graal.

Gomes, Tiago de Melo (2004) *Espelho no palco. Identidades sociais e massificação da cultura no teatro de revista dos anos 1920*. Campinas: Ed. da Unicamp.

Gómez-Peña, Guillermo (2005) *Ethno-techno: Writings on Performance, Activism, and Pedagogy*. New York: Routledge.

Gonçalves, José Moura Filho (1988) "Olhar e memória." In Adauto Novaes (ed.), *O Olhar*. São Paulo: Companhia das Letras.

Gonsalves da Silva, Vagner (2004) *Memória Afro-Brasileira 2: Artes do Corpo*. São Paulo: Selo Negro.

Gonzalez, Lélia (1985) "The Unifíed Black Movement: A New Stage in Black Political Mobilization." In Pierre-Michel Fontaine (ed.), *Race, Class and Power in Brazil*. Los Angeles: UCLA.

Gottschild, Brenda Dixon (1998) *Digging the Africanist Presence in American Performance: Dance and Other Contexts*. Westport: Praeger.

Greiner, Christine (1998) *Etnocenologia. Textos selecionados,*. São Paulo: Annablume.

——. (2006) *O corpo. Pistas para estudos indisciplinares*, 2nd edn. São Paulo: Annablume.

——. (2007) *Leituras do corpo*. São Paulo: Annablume.

——. (2010) *O corpo em crise. Novas pistas e o curto-circuito das representações*. São Paulo: Annablume.

Greskovic, R (2000) *Ballet: A Complete Guide*. London: Robert Hale.

——. (2005) *Ballet 101: A Complete Guide to Learning and Loving the Ballet*. New York: Limelight Editions.

Grinker, Roy (1997) *Perspectives on Africa: A Reader in Culture, History, and Representation*. Oxford: Blackwell.

Grosz, E. (1994) *Volatile Bodies: Toward a Corporeal Feminism*. Bloomington: Indiana University Press.

Guillermoprieto, Alma (1991) *Samba*. New York: Vintage Books.

Hall, Stuart (1992) "The Question of Cultural Identity." In D. Hall, S. Held, and T. McGrew (eds), *Modernity and its Futures*. Cambridge: Polity Press.

——. (2003) *A identidade cultural na pós-modernidade*. Rio de Janeiro: DP&A Ed.

Hartman, Saidiya V (1997) *Scenes of Subjection: Terror, Slavery, and Self-Making in Nineteenth-Century America*. Oxford University Press.

Heidegger, Martin (1962) *Being and Time*. John Macquarrie and Edward Robinson (trans.). London: SCM Press.

Hewitt, Andrew (*2005*) *Social Choreography: Ideology as Performance in Dance and Everyday Movement*. Durham, NC: Duke University Press.

Hobsbawn, Eric (1997) *A invencão das tradicões*. Rio de Janeiro: Paz e Terra.

Holanda, Sérgio Buarque de (2006) *Raízes do Brasil. Edição comemorativa de 70 anos*, Ricardo Benzaquen de Araújo and Lilia Moritz Schwarcz (eds). São Paulo: Companhia das Letras.

Howson, Alexandra (2005) *Embodying Gender*. Thousand Oaks, CA: Sage.

Huizinga, Johan (1955) *Homo Ludens: A Study of the Play-Element in Culture*. Boston: Beacon Press.

Jacques, Paola Berenstern (2001) *Estética da Ginga. A arquitetura das favelas através da obra de Hélio Oiticia*. Rio de Janeiro: Ed. Casa da palavra.

Jameson, Fredric (1995) *Marcas do visível*. Rio de Janeiro: Graal.

Jancsó, István and Kantor Iris (2001) *Festa, cultura e sociabilidade na América portuguesa*. São Paulo: EDUSP.

——. (eds) (2001) *Seminário Internacional "Festa: cultura e sociabilidade na América portuguesa."* São Paulo: Hucitec; Edusp; FAPESP; Imprensa Oficial SP.

Joaquim, Maria (2001) *O papel da liderança religiosa feminina na construção da identidade negra*. Rio de Janeiro: Pallas.

Johnson, E. Patrick (2003) *Appropriating Blackness: Performance and the Politics of Authenticity*. Durham, NC: Duke University Press.

Jório, Amaury and Araujo, Hiram (1969) *Escolas de samba em desfile, vida, paixão e sorte*. Rio de Janeiro: Poligráf.

Julião, Carlos (1960) *Riscos illuminados de figurinhos de brancos e negros dos uzos do Rio de Janeiro e Serro do Frio, aquarelas de Carlos Julião*, Lygia da Fonseca Fernandes da Cunha (ed.). Rio de Janeiro: Livraria São José.

Kabengele, Munanga (1999) *Rediscutindo a mestiçagem no Brasil: identidade nacional versus identidade negra*. Petrópolis: Editora Vozes.

Kagame, A (1975) "A Percepção Empírica do Tempo e Concepção da História no Pensamento Bantu." In R. Ricouer (ed.), *As culturas e o tempo: estudos reunidos pela UNESCO*. Rio de Janeiro: Vozes.

Karasch, Mary C. (1987) *Slave Life in Rio de Janeiro, 1808–1850*. Princeton University Press.

Katz, Helena (1989) *Danças populares brasileiras*. São Paulo: Rhodia.

——. (1994) *O Brasil descobre a dança (Brazil Discovers the Dance)*. São Paulo: Dórea Books and Art.

——. (1995) *Grupo Corpo: Brazilian Dance Theatre*. Rio de Janeiro: Salamandra.

——. (2003) "Dança, o Pensamento do Corpo." In A. Novaes (ed.), *O Homem-Maquina*. São Paulo: Companhia das Letras.

——. (2005) *Um, dois, três, a dança é o pensamento do corpo*. Belo Horizonte: FID.

——. (2006) "O corpo e o Meme Laban: uma trajetória evolutiva." In M. Mommensohn and P. Petrella (ed.), *Reflexões sobre Laban, o mestre do movimento*. São Paulo: Summus.

Kaz, Leonel (2003) *Brasil, rito e ritmo. Um século de música popular e clássica*. Rio de Janeiro: Aprazivel Edições.

Kirshenblatt-Gimblett, Barbara (1998) *Destination Culture: Tourism, Museums, and Heritage*. Berkeley: University of California Press.

Kraay, Hendrik (1998) *Afro-Brazilian Culture and Politics: Bahia, 1790s to 1990s*. Armonk: M.E. Sharpe.

Kraut, A (2008) *Choreographing the Folk: The Dance Stagings of Zora Neale Hurston*. Minneapolis: University of Minnesota Press.

Kubik, Gerhard (1979) *Angolan Traits in Black Music, Games and Dances of Brazil: A Study of African Cultural Extensions Overseas*. Lisbon: Junta de Investigações Científicas do Ultramar.

Lafit, Miran de Barros (1965) *Uma cidade no trópico – São Sebastião do Rio de Janeiro*, 2nd edn. Rio de Janeiro: AGIR.

Leitch, Vincent (2001) *The Norton Anthology of Theory and Criticism*. New York: Norton.

Leopoldi, José Sávio (1978) *Escola de Samba, ritual e sociedade*. Petrópolis: Vozes.

Lepecki, André (2006) *Exhausting Dance: Performance and the Politics of Movement*. New York: Routledge.

Levi, Clovis (1997) *Teatro brasileiro, um panorama do século XX*. Rio de Janeiro: FUNARTE.

Lewis, John Lowell (1992) *Ring of Liberation: Deceptive Discourse in Brazilian Capoeira*. University of Chicago Press.

Ligiéro, Zeca (2004) *Malandro Divino: a vida e a lenda de Zé Pelintra personagem mítico da lapa carioca*. Rio de Janeiro: Ed. Record.

Lima, Evani Tavares (2002) *Capoeira Angola como treinamento para o ator*. Salvador: UFBA.

Lima, Valéria (2004) *Viagem com Debret*. Rio de Janeiro: Zahar.

Lobo, Lenora (2003) *Teatro do movimento: um método para o intérprete criador*. Brasília: LGE.

Lody, Raul (1988) *Pencas de balangandãs da Bahia. Um estudo etnográfico das jóias-amuletos*. Rio de Janeiro: FUNARTE.

Lopes, André Luiz Lacé. (2002) *A capoeiragem no Rio de Janeiro: primeiro ensaio, Sinhozinho e Rudolf Hermanny*. Rio de Janeiro: Editora Europa.

——. (1999) *A volta do mundo da capoeira*. Rio de Janeiro: Coreográfica.

Lopes, Antônio Herculano (2000) *Entre Europa e Africa, a invenção do carioca*. Rio de Janeiro: Topbooks.

Lopes, Nei (1988) *Bantos, malês, e a identidade negra*. Rio de Janeiro: Forense Universitária,

——. (1992) *O negro no Rio de Janeiro e sua tradição musical. Partido-alto, calango, chula e outras cantorias*. Rio de Janeiro: Pallas.

——. (1999) *171-Lapa-Irajá: casos e enredos do samba*. Rio de Janeiro: Edições Folha Seca.

——. (2003a) *Novo dicionário banto do Brasil, contendo mais de 250 propostas etmológicas acolhidas pelo dicionário Houaiss*. Rio de Janeiro: Pallas.

——. (2003b) *Sambeabá: o samba que não se aprende na escola*. Rio de Janeiro: Casa da Palavra.

——. (2005) *Partido-Alto, samba de bamba*. Rio de Janeiro: Pallas.

Macedo, Tânia (2006) "Angola e Brasil: com os malandros em cena." In *Brasil/ África, como se o mar fosse mentira*. São Paulo: UNESP.

Machado de Assis, Joaquim M. (1881). *Memórias Póstumas de Brás Cubas*. São Paulo: Atica.

——. (1996) *Terpsícore*. São Paulo: Boitempo Editorial.

——. (1998) *Balas de estalo*, Heloisa Helena Paiva de Luca (ed.). São Paulo: Annablume.

Machado de Assis, Joaquim M. and Duarte, Eduardo de Assis (2007) *Machado de Assis afro-descendente, escritos de caramujo*. Rio de Janeiro: Pallas.

Machado, Cacá (2007) *O enigma do homem célebre: ambição e vocação de Ernesto Nazareth*. São Paulo: Instituto Moreira Salles.

Magaldi, Ana Maria Bandeira de Mello, Raposo, Figueiredo, and Raposo, Luciano (1971) "Quitandas e quitutes; um estudo sobre a rebeldia e transgressão femininas numa sociedade colonial." In *Cadernos de Pesquisa*. São Paulo: Fundação Carlos Chagas.

Maior, Mário Souto and Valente, Waldemar (1988) *Antologia pernambucana de folclore*. Recife: Fundação Joaquim Nabuco, Editora Massangana.

Manning, Susan (2004) *Modern Dance, Negro Dance: Race in Motion*. Minneapolis: University of Minnesota Press.

——. (2006) *Ecstasy and the Demon: The Dances of Mary Wigman*. Minneapolis: University of Minnesota Press.

Marcuse, Herbert (1974) *Eros and Civilization: A Philosophical Inquiry into Freud*. Boston: Beacon.

Marques, Isabel (1999) *Ensino de Dança Hoje*. São Paulo: Cortez.

Martin, Randy (1998) *Critical Moves: Dance Studies in Theory and Politics*. Durham, NC: Duke University Press.

Martín-Barbero, Jesús (1987) *De los medios a las mediaciones*. Barcelona: Gustavo Gili.

——. (2003) *Dos meios às mediações*. Rio de Janeiro: UFRJ.

Martins, Suzana (2008) *A dança de Yemanjá Oguntê sob a perspectiva estética do corpo*. Salvador: S. Martins.

Matory, J. Lorand (2006) "Tradition, Transnationalism and Gender in the Afro-Brazilian Candomble." In Doris Sommer (ed.), *Cultural Agency in the Americas*. Durham, NC: Duke University Press.

Mattos, Regiane (2007) *História e cultura afro-brasileira*. São Paulo: Contexto.

McGowan, Chris and Ricardo Pessanha (1998) *The Brazilian Sound: Samba, Bossa Nova, and the Popular Music of Brazil*. Philadelphia: Temple University Press.

McNally, David (2001) *Bodies of Meaning: Studies on Language, Labor, and Liberation*. Albany: State University of New York Press.

Meireles, Cecília (1983) *Batuque, samba and macumba. Estudos de gesto e ritmo, 1926–1934*. Rio de Janeiro: Funarte, INF.

Mello, Luiz Gonzaga de (1990) *Pastoril profano de Pernambuco*. Recife: Fundação Joaquim Nabuco, Editora Massangana.

Merleau-Ponty, Maurice (2008) *The Visible and the Invisible*. Princeton: Recording for Blind & Dyslexic.

Merrell, Floyd (2005) *Capoeira and Candomblé: Conformity and Resistance in Brazil*. Princeton: Markus Wiener Publishers.

Michailowsky, Pierre (1956) *A dança e a escola de ballet*. Rio de Janeiro: Ministério da Educação e Cultura.

Mignolo, Walter (2005) *The Idea of Latin America*. Oxford: Blackwell.

Mitchell, Michael (1985) "Blacks and the abertura democrática." In Pierre-Michel Fontaine (ed.), *Race, Class and Power in Brazil*. Los Angeles: UCLA.

Mohanram, Radhika (1999) *Black Body: Women, Colonialism, and Space*. Minneapolis: University of Minnesota Press.

Mommensohn, Maria and Petrella, Paulo (2006) *Reflexões sobre Laban, o mestre do movimento*. São Paulo: Summus.

Moraes, Vinícius (2008) *Samba falado: crônicas musicais*. Rio de Janeiro: Azougue Editorial.

Moraña, M., Dussel, E.D., and Jáuregui, C.A. (eds) (2008). *Coloniality at Large: Latin America and the Postcolonial Debate*. Durham, NC: Duke University Press.

Moura, Carlos (2000) *A travessia da Calunga Grande: três séculos de imagens sobre o negro no Brasil, 1637–1899*. São Paulo Brasil: Edusp.

Moura, Jair (1979) *Capoeira, a luta regional baiana*. Salvador: Prefeitura Municipal.
——. (1980) *Capoeiragem—arte and malandragem*. Salvador: Bureau.
——. (1993) *Mestre Bimba: a crônica da capoeiragem*. Salvador: Fundção Mestre Bimba.
Moura, Roberto (1983) *Tia Ciata e a pequena África no Rio de Janeiro*. Rio de Janeiro: FUNARTE.
——. (2004) *No principio, era o samba. Um estudo sobre samba, partido-alto e outros pagodes*. Rio de Janeiro: Rocco.
Munanga, Kabengele (2004) *Rediscutindo a mestiçagem no Brasil: Identidade Nacional versus identidade negra*. Belo Horizonte: Autêntica.
Murphy, John Patrick (2006) *Music in Brazil*. New York: Oxford University Press.
Naro, Nancy (2007) *Cultures of the Lusophone Black Atlantic*. New York: Palgrave Macmillan.
Nascimento, Abdias do (1992) *Africans in Brazil: A Pan-African Perspective*. Trenton, NJ: Africa World Press.
——. (2002) *O Brasil na Mira do Pan-africanismo*, 2nd edn. Salvador: Edufba, CEAO.
Nasser, David (1966) *A Vida trepidante de Carmem Miranda*. Rio de Janeiro: O Cruzeiro.
Navas, Cássia (1999) *Dança e mundialização: políticas de cultura no eixo Brasil-França*. São Paulo: Hucitec.
——. (ed.) (2003) *Balé da Cidade de São Paulo*, Camilo Rocha (trans.). São Paulo: Formarte.
Navas, Cássia and Dias, Linneu (1992) *Dança moderna*. São Paulo: Secretaria Municipal de Cultura.
Ness, Sally Ann (1992) *Body, Movement, and Culture: Kinesthetic and Visual Symbolism in a Philippine Community*. Philadelphia: University of Pennsylvania Press.
Novack, Cynthia (1990) *Sharing the Dance: Contact Improvisation and American Culture*. Madison: University of Wisconsin Press.
Nogueira, N. *et al.* (eds) (2005) *Dossiê das Matrizes do Samba no Rio de Janeiro*. Brasília: IPHAN.
Novaes, Adauto (2003) *O homem-máquina: a ciência manipula o corpo*. São Paulo: Companhia das Letras.
Ochoa, Enrique (2005) *Latino Los Angeles: Transformations, Communities, and Activism*. Tucson: University of Arizona Press.
Oliveira, Josivaldo (2005) *No tempo dos valentes: os capoeiras na cidade da Bahia*. Salvador: Quarteto.
Oliveira, Valdemar de (1971) *Frevo, Capoeira e "Passo."* Recife: Companhia Editora de Pernambuco.
——. (1977) *O capoeira, um teatro do passado*. Brasilia: Departamento de Documentação.
Ortiz, Fernando (1995) *Cuban Counterpoint, Tobacco and Sugar*. Durham, NC: Duke University Press.
Ortiz, Renato (1985) *Cultura brasileira e identidade nacional*. São Paulo: Brasiliense.
——. (1988) *Morte branca de um feiticeiro negro: umbanda e sociedade brasileira*. São Paulo: Brasiliense.

O'Shea, Janet (2007) *At Home in the World: Bharata Natyam on the Global Stage*. Middletown, CT: Wesleyan University Press.
——. (2010) "Roots/Routes of Dance Studies." In A. Carter and J. O'Shea (eds), *The Routledge Dance Studies Reader*, 2nd edn. New York: Routledge.
Otero, Décio (1999) *Stagium, as paixões da dança*. São Paulo: Hucitec.
Paezzo, Sylvan (1972) *Memórias de Madame Satã: Conforme Narração a Sylvan Paezzo*. Rio de Janeiro: Lidador.
Paiva, Salvyano Cavalcanti de (1991) *Viva o rebolado! Vida e morte do teatro de revista brasileiro*. Rio de Janeiro: Editora Nova Fronteira.
Pantoja, Selma (2000) *Nzinga Mbandi: mulher, guerra e escravidão*. Brasilia: Thesaurus.
——. (ed.) (2001) *Entre Áfricas e Brasis*. Brasilia: São Paulo Paralelo 15, Marco Zero.
Pantoja, Selma and Saraiva, José Flávio (1999) *Angola e Brasil nas Rotas do Atlântico Sul*. Rio de Janeiro: Bertrand Brasil.
Pastinha, Vicente Ferreira (1964) *Capoeira Angola*. Salvador: Loreto.
Pastinha, Vicente Ferreira (1996) *Manuscritos e desenhos de Mestre Pastinha*, Angelo Decanio Filho (ed.) Salvador. Available at: http://portalcapoeira.com/downloads-capoeira/manuscritos-e-biografias/55-os-manuscritos-do-mestre-pastinha (date accessed May 23, 2015).
Pereira, Marcelo (1990) *Arturos, olhos do rosário*. Belo Horizonte: Mazza.
Pereira, Roberto (2003) *A formação do balé brasileiro: nacionalismo e estilização*. Rio de Janeiro: FGV Editora.
——. (2004) *Eros Volusia, a criadora do bailado nacional*. Rio de Janeiro: Relume Dumará.
Perna, Marco Antonio (2005) *Samba de gafieira. A história da dança de salão brasileira*, 2nd edn. Rio de Janeiro: self-published.
Phelan, Peggy (1993) *Unmarked: The Politics of Performance*. New York: Routledge.
Pires, Antônio Liberac Cardoso Simões (2004) *A Capoeira na Bahia de Todos os Santos. Um estudo sobre cultura e classes trabalhadoras (1890–1937)*. Porto Nacional: Fundação Federal do Tocantins, NEAB; Grafset.
Pollock, Della (1998) *Exceptional Spaces: Essays in Performance and History*. Chapel Hill: University of North Carolina Press.
Powe, Edward L. (2002) *Danced Martial Arts of the Americas*. Madison: Dan Aiki Publications.
Prandi, Reginaldo (1991) *Os candomblés de S. Paulo*. São Paulo: Hucitec.
Querino, Manuel Raymundo (1909) *As artes na Bahia*. Rio de Janeiro: Imprensa Nacional.
——. (1914) *Bailes Pastoris*. Salvador: Typographia Almeida.
——. (1916) *A Bahia de Outrora*. Salvador: Livraria Econômica.
——. (1955) *A raça africana*. Salvador: Progresso.
——. (1978) *The African Contribution to Brazilian Civilization*. Tempe: Arizona State University, Center for Latin American Studies.
——. (1988) *Costumes Africanos no Brasil*. Recife: Editora Massangana.
Ramos, Arthur (1939) *The Negro in Brazil*. Washington DC: Associated Publishers.
Ramsey, Kate (1997) "Voudou, Nationalism, and Performance: The Staging of Folklore in Mid-Twentieth Century Haiti." In Jane C. Desmond (ed.), *Meaning and Motion: New Cultural Studies of Dance*. Durham, NC: Duke University Press.
Rego, José Carlos (1996) *Dança do samba: Exercício do prazer*. Rio de Janeiro: Aldeia.

Rego, Valdeloir (1968) *Capoeira Angola: ensaio sócio-etnográfico*. Salvador: Editora Itapuã.
Reis, João José (1993) *Slave Rebellion in Brazil: The Muslim Uprising of 1835 in Bahia*. Baltimore: Johns Hopkins University Press.
——. (2003) *Rebelião escrava no Brasil: a história do levante dos malês em 1835*. São Paulo: Companhia das Letras.
Reis, Letícia (1997) *O mundo de pernas para o ar, a capoeira no Brasil*. São Paulo: Publisher Brasil.
Renan, Ernest (1990) "What is a Nation?" In Homi K. Bhabha (ed.), *Nation and Narration*. New York: Routledge.
Renault, Delso (1982) *O dia-a-dia no Rio de Janeiro segundo os Jornais 1870–1889*. Rio de Janeiro: Civilização Brasileira.
Ribeiro, André (1999) *O diamante eterno: biografia de Leônidas da Silva*. Rio de Janeiro: Gryphus.
Rio, João do (1987) *A alma encantadora das ruas*. Rio de Janeiro: Sec. Mun. de Cultura.
Roach, Joseph R. (1996) *Cities of the Dead: Circum-Atlantic Performance*. New York: Columbia University Press.
——. (1998) "Slave Spectacles and Tragic Octoroons." In Delia Pollock (ed.), *Exceptional Spaces: Essays in Performance and History*. Chapel Hill: University of North Carolina Press.
——. (2003) "Culture and Performance in the Circum-Atlantic World." In E. Striff (ed.), *Performance Studies*. New York: Palgrave Macmillan.
Robatto, Lia (1994) *Dança em processo: a linguagem do indizível*. Salvador: Centro Editorial e Didático da UFBA.
——. (2002) *Passos da dança - Bahia*. Salvador: Fundação Casa de Jorge Amado.
Rocha, José Maria Tenório (1984) *Folguedos e danças de Alagoas: Sistematização e Classificação*. Maceió: Secretaria de Educação e Cultura de Alagoas, Departamento de Assuntos Culturais, Comissão Alagoana de Folclore.
Rocha, Oswaldo Porto (1986) *A era das demolições: a cidade do Rio de Janeiro 1870–1920*. Rio de Janeiro, Secretaria Mun. de Cultura.
Rodrigues, Graziela (1997) *Bailarino, Pesquisador, Interprete: processo de formação*. Brasilia: Funarte.
Rodrigues, J.C (1986) *Tabu do corpo*, 4th edn. Rio de Janeiro: Dois Pontos.
Rodrigues, Nina (1988) *Os africanos no Brasil*. Brasilia: Editora da UnB.
Rodrigues, Raymundo Nina (1932) *Os Africanos no Brasil*. São Paulo: Editora Nacional.
Rodrigues Filho, Mário (1947) *O negro no football brasileiro*. Irmãos Pongetti Editores: Rio de Janeiro.
——. (2003) *O negro no futebol brasileiro*. Rio de Janeiro: Mauad.
Rose, Nikolas (1999) *Powers of Freedom: Reframing Political Thought*. Cambridge University Press.
Rose, Tricia (1994) *Black Noise: Rap Music and Black Culture in Contemporary America*. Hanover, NH: Wesleyan University Press.
——. (1996) "A Style Nobody Can Deal with: Politics, Style, and the Postindustrial City in Hip Hop." In A.F. Gordon and C. Newfield (eds), *Mapping Multiculturalism*. Minneapolis: University of Minnesota Press.
Rosenfeld, Anatol (1993) *Negro, macumba e futebol*. São Paulo: Perspectiva/ EDUSP.

Rubim, Antonio Albino Canelas, "Políticas culturais no Brasil: trajetória e contemporaneidade" In *Lei Rouanet Percursos e relatos de 2011*, available at: www.atitudebrasil.com/site/wp-content/uploads/2011/10/rouanet_web.pdf (date accessed May 11, 2015).

Rugendas, Johann Moritz (1835) *Voyage Pittoresque dans le Bresil*, Marie Philippe Aimé de Golbery (trans.). Paris: Engelmann.

——. (1967) *Viagem Pitoresca através do Brasil*, Sérgio Milliet (trans.). São Paulo: Livaria Martins.

Sade, Marquise de (1976) *La philosophie dans le boudoir: ou Les instituteurs immoraux*. Paris: Gallimard.

Said, Edward (1979) *Orientalism*. New York: Vintage Books.

——. (1996) *Representations of the Intellectual: The 1993 Reith Lectures*. New York: Vintage Books.

Sandroni, C. (2001) *Feitiço decente*. Rio de Janeiro: UFRJ: Zahar.

Sandroni, C. *et al.* (eds) (2004) *Samba de roda do Recôncavo Baiano*. Brasilia: IPHAN.

Sansone, Livio and Santos, Jocélio Teles dos (eds) (1998) *Rítmos em transito. Socio-antropologia da música na Bahia*. São Paulo: Dynamis.

Sansone, Livio (2004) *Negritude sem etnicidade: o local e o global nas relações raciais e na produção cultural negra do Brasil*. Salvador/Rio de Janeiro: Edufba/Pallas.

Sant'Anna, Affonso Romano de (1984) *Canibalismo amoroso: o desejo e a interdição em nossa cultura através da poesia*. São Paulo: Brasiliense.

Santos, Inaicyra Falcão dos (2002) *Corpo e ancestralidade: uma proposta pluricultural de dança-arte-educação*. Salvador: EDUFBA.

Santos, Jocélio (2005) *O poder da cultura e a cultura no poder: disputa simbólica da herança cultural negra no Brasil*. Salvador: EDUFBA.

Santos, Myrian S. dos (1999) "Samba Schools: The Logic of Orgy and Blackness in Rio de Janeiro." In Jean M. Rahier (ed.), *Representations of blackness and the performance of identities*. Westport: Bergin & Garvey.

Sarmento, Alfredo de (1880) *Os Sertões D'Africa (Apontamentos de Viagem)*. Lisbon: Francisco Arthur da Silva.

Savigliano, Marta (1995) "Tango and the Postmodern Uses of Passion." In Sue-Ellen Case, Philip Brett, and Susan Leigh Foster (eds), *Cruising the Performative: Interventions into the Representation of Ethnicity, Nationality, and Sexuality*. Bloomington: Indiana University Press.

——. (1995) *Tango and the Political Economy of Passion*. Boulder: Westview Press.

——. (2011). "Worlding Dance and Dancing Out There in the World." In S.L. Foster (ed.), *Worlding Dance*. Basingstoke: Palgrave Macmillan.

Schechner, Richard (1976) *Ritual, Play, and Performance: Readings in the Social Sciences Theatre*. New York: Seabury Press.

Schwarcz, Lília M. (1993) *O Espetáculo das raças. Cientistas, instituições e questão racial no Brasil. 1870–1930*. São Paulo: Companhia das Letras.

——. (2001) *Racismo no Brasil*. São Paulo: Publifolha.

Schwarcz, Lilia M. and Reis, Letícia V. de S. (eds) (1996) *Negras imagens. Ensaios sobre cultura e escravidão no Brasil*. São Paulo: EdUSP.

Senna, Carlos (1990) *Capoeira, percurso*. Salvador: Senavox and A Rasteira.

Shaw, Lisa (1999) The *Social History of the Brazilian Samba*, 2nd edn. Burlington: Ashgate.

Shea Murphy, J. (2007) *The People Have Never Stopped Dancing: Native American Modern Dance Histories*. Minneapolis: University of Minnesota Press.

Shepherd, Gregory (2006) *Communication as ...: Perspectives on Theory*. Thousand Oaks, CA: Sage.

Silva, Ana Márcia and Damiani, Iara Regina (eds) (2005) *Praticas corporais, Experiências em Educação física para a outra formação humana*, vol. 3. Florianópolis: UFSC.

——. (2006) *Praticas corporais, construindo outro saberes em educação física*, vol. 2. Florianópolis: UFSC.

Silva, Eliana (2004) *Dança e pós-modernidade*. Salvador: EDUFBA.

Silva, Gladson (2008) *Capoeira: um instrumento psicomotor para a cidadania*. São Paulo: Phorte Editora.

Silva, Jose Milton Ferreira da (2004) *A linguagem do corpo na capoeira*. Rio de Janeiro: Sprint.

Silva, Soraia (2007) *Poemadançando: Gilka Machado e Eros Volúsia*. Brasilia: Editora UnB.

Siqueira, Maria de Lourdes (2006) *Imagens negras: Ancestralidade, diversidade e educação*. Belo Horizonte: Maza.

Skidmore, Thomas (1989) *Preto no Branco, raça e nacionalidade no pensamento brasileiro*. Rio de Janeiro: Paz e Terra.

Smiers, Joost (2003) *Arts under Pressure: Promoting Cultural Diversity in the Age of Globalization*. London: Zed Books.

Soares, Carlos Eugênio Líbano (1994) *A negregada instituição. Os capoeiras no Rio de Janeiro, 1850–1890*. Rio de Janeiro: Secretaria Municipal de Cultura.

——. (2004) *A capoeira escrava e outras tradições rebeldes no Rio de Janeiro, 1808–1850*, 2nd edn. Campinas: Editora Unicamp.

Sodré, Muniz (1988) *Santugri, histórias de mandinga e capoeiragem*. Rio de Janeiro: Rocco.

——. (1991) *O Brasil Simulado e o real: Ensaios sobre o cotidiano nacional*. Rio de Janeiro: Rio fundo.

——. (1998) *Samba, o dono do corpo*, 2nd edn. Rio de Janeiro: Mauad.

——. (1999) *Claros e escuros: identidade, povo e mídia no Brasil*. Petrópolis: Editora Vozes.

——. (2002a) *Mestre Bimba: Corpo e Mandinga*. Rio de Janeiro: Manati.

——. (2002b) *O terreiro e a cidade: a forma social negro-brasileira*. Rio de Janeiro: Imago Ed.

——. (2005) *A verdade seduzida, por um conceito de cultura no Brasil*, 3rd edn. Rio de Janeiro: DP&A.

Sommer, Doris (2006) *Cultural Agency in the Americas*. Durham, NC: Duke University Press.

Spinoza, B. (1993 [1677]). *Ethics and Treatise on the Correction of the Intellect*, A. Boyle (trans.). London: J.M. Dent.

Spivak, Gayatri Chakravorty (1988) "Can the Subaltern Speak?" In Cary Nelson and Larry Grossberg (eds), *Marxism and the Interpretation of Culture*. Chicago: University of Illinois Press.

——. (1996) *Spivak Reader: Selected Works of Gayatri Chakravorty Spivak*, Donna Landry and Gerald MacLean (eds). New York: Routledge.

——. (1999) *A Critique of Postcolonial Reason: Toward a History of the Vanishing Present*. Cambridge, MA: Harvard University Press.

Strazzacappa, Márcia (2003) "Reflexões sobre a formação profissional do artista da dança." In *Lições da dança*, vol. 4. Rio de Janeiro: UniverCidade.
Sucena, Eduardo (1989) *A Dança Teatral no Brasil*. Rio de Janeiro: Editora Memória.
Sweeney, Sean (2002) *The Body*. Cambridge University Press.
Tait, Luis (1996) *O Cancionista: Composição de Caçoes no Brasil*. São Paulo: EdUsp.
Talmon-Chvaicer, Maya (2008) *The Hidden History of Capoeira: A Collision of Cultures in the Brazilian Battle Dance*. Austin: University of Texas Press.
Tavares, Luiz Carlos (2006) *O Corpo que ginga, joga e luta. A Corporeidade na Capoeira*. Salvador: self-published.
Taylor, Diana (2003) *The Archive and the Repertoire: Performing Cultural Memory in the Americas*. Durham, NC: Duke University Press.
Taylor, Gerard (2005) *Capoeira*. Berkeley: North Atlantic Books.
Teles dos Santos, Jocélio (2005) *O poder da cultura e a cultura do poder. A disputa simbólica da herança cultural negra no Brasil*. Salvador: Edufba.
Thompson, Robert and Joseph Cornet (1981) *The Four Moments of the Sun: Kongo Art in Two Worlds*. Washington DC: National Gallery of Art.
Thompson, Robert Farris (1983) *Flash of the Spirit: African and Afro-American Art and Philosophy*. New York: Random House.
——. (1991) *Dancing between Two Worlds: Kongo-Angola Culture and the Americas*. New York: Caribbean Cultural Centre.
——. (2002) *Geste kôngo: exposition, Paris, Musée Dapper, 18 septembre 2002–19 janvier 2003*. Paris: Musée Dapper.
Tinhorão, José Ramos (1988a) *Os sons dos negros no Brasil: cantos, danças, folguedos - origens*. São Paulo: Art.
——. (1988b) *História social da música popular brasileira*. São Paulo: Editora 34.
——. (1991) *Pequena história da música popular. Da modinha à lambada*, 6th edn. Rio de Janeiro: Art Editora.
——. (2000) *As festas no Brasil colonial*. São Paulo: Editora 34.
——. (2001) *Cultura popular: temas e questões*. São Paulo: Editora 34.
——. (2006) *O rasga, uma dança negro-portuguesa*. São Paulo: Editora 34.
Tramonte, Cristiana (2001) *O samba conquista passagem. As estratégias e a ação educativa das escolas de samba*. Petrópolis: Editora Vozes.
Travasso, Elizabeth (2007) *Jongo do Sudeste*. Brasilia: IPHAN.
Treece, David (2000) "Rhythm and Poetry: Politics, Aesthetics and Popular Music in Brazil since 1960." In Anny Brooksbank Jones and Ronaldo Munck (eds), *Cultural Politics in Latin America*. Basingstoke: Macmillan.
Trigo, Luciano (2001) *O viajante imóvel, Machado de Assis e o Rio de Janeiro de seu tempo*. Rio de Janeiro: Editora Record.
Turner, Victor (2001) *From Ritual to Theatre: The Human Seriousness of Play*. New York: PAJ Publications.
Ulloa, Alejandro (1998) *Pagode, a festa do samba no Rio de Janeiro e nas Américas*. Rio de Janeiro: MultiMais Editorial.
Velloso, Mônica Pimenta (1985) *A brasilidade verde-amarela: nacionalismo e regionalismo paulista*. Rio de Janeiro: CPDOC.
——. (1988) *As tradições populares na "Belle Époque" carioca*. Rio de Janeiro: Funarte.
Veloso, Caetano (2001) "Carmen Miranda Dada." In Christopher Dunn and Charles A Perrone (eds), *Brazilian Popular Music and Globalization*. Gainesville: University Press of Florida.
——. (2002) *Tropical Truth: A Story of Music and Revolution in Brazil*. New York: Knopf.

Veneziano, Neyde. (1991) *Teatro de revista no Brasil: dramaturgia e convenções*. Campinas: Editora da UNICAMP: Pontes.
Verger, Pierre (1981) *Notícias da Bahia – 1850*. Salvador: Corrupio.
Viana, Klauss (2005) *A dança*, 3rd edn. São Paulo: Summus.
Vianna, Hermano (1988) *O mundo funk carioca*. Rio de Janeiro: Zahar.
——. (1999) *The Mystery of Samba: Popular Music and National Identity in Brazil*, 5th edn. Chapel Hill: University of North Carolina Press.
Vianna, Luiz Fernando (2004) *Geografia carioca do samba*. Rio de Janeiro: Casa da Palavra.
Vieira, Luis Renato (1995) *O jogo da Capoeira: Corpo e Cultura popular no Brasil*. Rio de Janeiro: Sprint.
Vicenzia, Ida (1997) *Dança no Brasil, História visual*. São Paulo: Ministério da Cultura, FUNARTE, Atração Produções Ilimitadas.
Vogt, Carlos and Fry, Peter (1996) *A África no Brasil*. Cafundó: Companhia das Letras.
Volusia, Eros (1939) *Dança brasileira, a criação do bailado brasileiro*. Rio de Janeiro: self-published.
——. (1983) *Eu e a dança*. Rio de Janeiro: Revista Continente Editorial LTDA.
Wacquant, Loïc (2006) *Body and Soul: Notebooks of an Apprentice Boxer*. Oxford University Press.
Waskul, Dennis (2006) *Body Embodiment: Symbolic Interaction and the Sociology of the Body*. Aldershot: Ashgate.
Weil, Pierre and Tompakow, Roland (1981) *O corpo fala. A linguagem silenciosa da comunicação não verbal*, 12th edn. Petrópolis: Vozes.
Welsh-Asante, Kariamu (ed.) (1996) *African Dance: An Artistic, Historical, and Philosophical Inquiry*. Trenton, NJ: Africa World Press.
——. (2001) "Commonalties in African Dance: An Aesthetic Foundation." In A. Dils and A.C. Albright (eds), *Moving History/Dancing Cultures: A Dance History Reader*. Middletown: Wesleyan University Press.
Wisnik, José Miguel (2004) *Sem Receita: Ensaios e Canções*, Arthur Nestrovski (ed.). São Paulo: Publifolha.
——. (2008a) *Machado Maxixe: o Caso Pestana*. São Paulo: Publifolha.
——. (2008b) *Veneno remédio, o futebol e o Brasil*. São Paulo: Companhia das Letras.
Worthen, W.B. (2004) "Disciplines of the Text: Sites of Performance" In H. Bial (ed.), *The Performance Studies Reader*. New York: Routledge.
Yarbro-Bejarano, Yvonne (2000) "Traveling Transgressions: Cubanidad in Performances by Carmelita Tropicana and Manga Gomes." In *Reading and Writing the Ambiente: Queer Sexualities in Latino, Latin American, and Spanish Culture*. Madison: University of Wisconsin Press.
Zaluar, Alba and Alvito, Marcos (1998) *Um século de favela*. Rio de Janeiro: Fundação Getulho Vargas.
Zetola, Bruno and Seixas, Andréa (2008) *Capoeira: Texts from Brazil*, vol. 14. São Paulo: Grafica Brasil.

Journal articles

Almeida, Tereza Virginia de (2003) "Corpos em tensão: feminino, masculino e barroco no espetáculo Bach." *Revista de Estudos Femininos*, 11(1), January–June.

Available at: www.scielo.br/scielo.php?script=sci_arttext&pid=S0104-026X2003000100008&lng=en&nrm=iso (date accessed May 11, 2015).

Andrade, Marilia de and Kátia Canton (1996) "Overview of Dance Research in Brazil." *Dance Research Journal*, 28(2), pp. 114–22.

Andrade, Oswald (1928) "Manifesto Atropofágico." *Revista de Antropofagia*, 1(1).

Arrizón, Alicia (2002) "Race-ing Performativity through Transculturation, Taste and the Mulata Body." *Theater Research International*, 27(2), pp. 136–52.

Augel, Moema Parente (2000) "A fala identitaria: teatro afro-brasileiro hoje." *Afro-Asia*, 24, pp. 291–323.

Barbosa, Maria José Somerlate (2005) "Capoeira, a gramática do corpo e a dança das palavras." *Luso-Brazilian Review*, 42(1), pp. 78–98.

Barke, Michael, Escasany, Tony and, O'Hare, Greg (2001) "Samba: A Metaphor for Rio's Favelas?" *Cities*, 18(4), pp. 259–70.

Béhague, Gerard (1980) "Improvisation in Latin American Musics." *Music Educators Journal*, 66(5), pp. 118–25. Available at: www.jstor.org/pss/3395789 (date accessed May 11, 2015).

Bião, Armindo (1998) "O obsceno em cena, ou o tchan na boquinha da garrafa." *Repertório Teatro & Dança*, 1(1), pp. 9–21.

Blanco Borelli, Melissa (2008) "'¿Y ahora qué vas a hacer, mulata?': Hip choreographies in the Mexican Cabaretera Film *Mulata* (1954)." *Women & Performance: A Journal of Feminist Theory*, 18(3), pp. 215–33.

Bretas, Marcos Luiz (*1989)* "Navalhas e Capoeiras: uma outra queda." *Ciência Hoje*, 10(59), pp. 56–64.

Browning, Barbara (2005) "Incessant Daily Negotiations: Bill T. Jones's Floating the Tongue." *The Drama Review*, 49(2), pp. 87–92.

——. (2007) "Conversations with Rosângela Silvestre and Steve Coleman." *Women & Performance: A Journal of Feminist Theory*, 17(2), pp. 171–84.

Butler, Judith (1993) "Critically Queer." *GLQ: A Journal of Lesbian and Gay Studies*, 1(1), pp. 17–32.

Caillois, Roger and Halperin, Elaine P. (1955) "The Structure and Classification of Games." *Diogenes*, 3, pp. 62–75. Available at: http://dio.sagepub.com/cgi/reprint/3/12/62 (date accessed May 11, 2015).

Candido, Antonio (1970) "Dialética da malandragem: Caracterização das Memórias de um Sargento de Milícias." *Revista do Instituto de Estudos Brasileiros*, 8, pp. 67–89.

Carvajal, Farmer. "The Experimental Exercise of Freedom: A Conversation with Rina Carvajal and Alma Ruiz John Alan Farmer." *Art Journal*, 59(1), pp. 22–31.

——. (1988) "Mestiçagem e Segregação." *Revista Humanidades*, V(17), pp. 35–9.

——. (1990) "O Jogo das Bolinhas: Uma Simbólica da Masculinidade." *Anuário Antropológico*, 87, pp. 191–222.

Carvalho, José Jorge de (1994) "The Multiplicity of Black Identities in Brazilian Popular Music." *Série Antropologia*, 163.

——. (1996) "Images of the Black Man in Brazilian Popular Culture." *Série Antropoligia*, 201.

——. (1999) "Afro-Brazilian Music and Ritual – Part 1." *Série Antropologia*, 256.

Castro Junior, Luís Vitor (2003) "Capoeira angola: olhares e toques cruzados entre historicidade e ancestralidade." *Revista Brasileira de Ciências do Esporte*, 25(2), pp. 143–58.

Catunda, Eunice (1952) "Capoeira no Terreiro de Mestre Waldemar." *Fundamentos - Revista de Cultura Moderna,* 30, pp. 16–18.

Cavalcanti, Maria Laura Viveiros de Castro (2002) "Os sentidos no espetáculo." *Revista Antropologia,* 45(1), pp. 37–78. Available at: www.scielo.br/scielo.php?script=sci_arttext&pid=S0034-77012002000100002&lng=en&nrm=iso (date accessed May 11, 2015).

Chasteen, John Charles (1996) "The Prehistory of Samba: Carnival Dancing in Rio de Janeiro, 1840–1917." *Journal of Latin American Studies,* 28(1), pp. 29–47. Available at: www.jstor.org/stable/157986 (date accessed May 11, 2015).

Cosentino, Donald (1987) "Who is That Fellow in the Many-Colored Cap? Transformations of Eshu in Old and New World Mythologies." *Journal of American Folklore,* 100(397), pp. 261–75.

Desch-Obi, T.J. (1992) "Capoeira: Martial Art as Spiritual Discipline." *Journal of Caribbean Studies,* Winter, pp. 87–98.

Desmond, Jane C. (1993–1994) "Embodying Difference: Issues in Dance and Cultural Studies." *Cultural Critique,* 26, pp. 33–63.

Dossar, Kenneth (1992) "Capoeira Angola: Dancing between Two Worlds." *Afro-Hispanic Review,* 11(1/3), pp. 5–10.

Downey, Greg (2002) "Listening to Capoeira: Phenomenology, Embodiment, and the Materiality of Music." *Ethnomusicology,* 46(3), pp. 487–509. Available at: http://links.jstor.org/sici?sici=0014-1836%28200223%2946%3A3%3C487%3ALTCPEA%3E2.0.CO%3B2-9 (date accessed May 11, 2015).

Drewal, H.J. and Drewal, M.T. (1987) "Composing Time and Space in Yoruba Art." *Word and Image: A Journal of Verbal/Visual Equiry,* 3(3), pp. 225–51.

Foster, Susan Leigh (1998) "Choreographies of Gender." *Signs,* 24(1), pp. 1–33.

——. (2002) "Walking and Other Choreographic Tactics: Danced Inventions of Theatricality and Performativity." *SubStance* 31(2–3), pp. 125–46.

Freire, Vanda Lima B (2004) "Operas e Magicas em Teatros e Saloes no Rio de Janeiro--Final do Seculo XIX, Inicio do Seculo XX." *Latin American Review,* 25(1), pp. 100–11.

Frigerio, Alejandro (1989) "Capoeira: de arte negra a esporte branco." *Revista Brasileira de Ciências Sociais,* 4(1), pp. 85–98.

Fry, Peter (2000) "Nationality, and the Meanings of "Race" in Brazil." *Daedalus,* 129(2), pp. 83–118. Available at: www.jstor.org/stable/20027630 (date accessed May 11, 2015).

Galinsky, Philip (1996) "Co-option, Cultural Resistance, and Afro-Brazilian Identity: A History of the 'Pagode' Samba Movement in Rio de Janeiro." *Latin American Music Review,* 17(2), pp. 120–49.

Gates, Jr., Henry Louis (1983) "The 'Blackness of Blackness': A Critique of the Sign and the Signifying Monkey." *Critical Inquiry,* 9(4), pp. 685–723.

Gilliam, Ângela (1998) "The Brazilian Mulata: Images in the Global Economy." *Race & Class,* 40(1), pp. 57–69.

Givan, B. (2007) "Apart Playing: McCoy Tyner and 'Bessie's Blues'."*Journal of the Society for American Music,* 1, pp. 257–80.

Gomes, Tiago de Melo (2004) "Para além da casa da Tia Ciata: outras experiências no universo cultural carioca, 1830–1930." *AfroÁsia,* 29–30, pp. 175–98.

Gomes, Tiago de Melo and Seigel, Micol (2002) "Sabina's Oranges: The Colours of Cultural Politics in Rio de Janeiro, 1889–1930." *Journal of Latin American Cultural Studies,* 11(1), pp. 5–28.

Green, James N. (2003) "O Pasquim e Madame Satã, a 'Rainha' Negra da Boemia Brasileira." *Topoi*, 4(7), pp. 201–21.

Holloway, Thomas H (1989) "A Healthy Terror: Police Repression of Capoeiras in Nineteenth-Century Rio de Janeiro." *Hispanic American Historical Review*, 69(4), pp. 637–76. Available at: www.jstor.org/stable/2516095 (date accessed May 11, 2015).

Jacques, Paola Berenstein (2001) "Estética das favelas." *Arquitextos*, 13, p. 8. Available at: www.vitruvius.com.br/revistas/read/arquitextos/02.013/883 (date accessed May 11, 2015).

Karel, Arnault (1988) "Africans Dance in Time: Kineasthetic Praxis and the Constructing of a Community." *Cultural Dynamics*, 1, pp. 252–81.

Katz, Helena (1999) "O coreógrafo como DJ." *Lições de dança*, 1, pp. 11–24.

——. (2000) "Os primeiros 25 anos deste Corpo." *Estudos Avançados*, 14(40), pp. 311–28. Available at: www.scielo.br/scielo.php?script=sci_arttext&pid=S0103-40142000000300023&lng=en&nrm=iso (date accessed May 11, 2015).

Katz, Helena and Greiner, Christine (2004) "O meio é a mensagem: porque o corpo é objeto da comunicação." *Húmus*, 1, pp. 11–19.

LaBoskey, Sara (2002) "Getting Off: Portrayals of Masculintiy in Hip Hop Dance." *Dance Research Journal*, 33(2), pp. 12–20.

Lara, Silvia Hunold (2002) "Customs and Costumes: Carlos Julião and the Image of Black Slaves in Late Eighteenth-Century Brazil." *Slavery & Abolition*, 23(2), pp. 123–46.

Lewis, J. Lowell (1995) "Genre and Embodiment: From Brazilian Capoeira to the Etymology of Human Movement." *Cultural Anthropology*, 10(2), pp. 221–43.

——. (1999) "Sex and Violence in Brazil: Carnaval, Capoeira, and the Problem of Everyday Life." *American Ethnologist*, 26(3), pp. 539–57.

Lobo, Lenora. "Teatro do movimento." *Humanidades*, 52, pp. 117–20.

Lugones, María and Price, Joshua (2003) "The Inseparability of Race, Class, and Gender in Latino Studies." *Latino Studies*, 1(2), pp. 329–32.

Maldonado-Torres, Nelson (2004) "The Topology of Being and the Geopolitics of Knowledge: Modernity, Empire, Coloniality." *City*, 8(1), pp. 29–56.

Maranhão, Tiago (2007) "Apollonians and Dionysians: The Role of Football in Gilberto Freyre's Vision of Brazilian People." *Soccer & Society*, 8(4), pp. 510–23.

Martin, Randy (2007) "Art Attacks." *Women & Performance: A Journal of Feminist Theory*, 17(3), pp. 283–97.

Marx, Anthony (1996) "Race-Making and the Nation-State." *World Politics*, 48(2), pp. 180–208.

Matta, Roberto da and Ray Green (1983) "An Interpretation of Carnaval." *SubStance*, 11(4), pp. 162–70.

Mauss, Marcel (1973) "Techniques of the Body." *Economy and Society*, 2(1), pp. 77–87.

McCann, Bryan (2001) "Noel Rosa's Nationalist Logic." *Luso-Brazilian Review*, 38(1), pp. 1–16.

Mignolo, Walter D. (1994) "Are Subaltern Studies Postmodern or Postcolonial? The Politics and Sensibilities of Geo-cultural Locations." *American Journal of Cultural Histories and Theories*, 9(46), pp. 45–73.

——. (2002) "The Geopolitics of Knowledge and the Colonial Difference." *South Atlantic Quarterly*, 101(1), pp. 57–96.

Muteba Rahier, Jean (1998) "Blackness, the Racial/Spatial Order, Migrations, and Miss Ecuador 1995–1996." *American Anthropologist*, 100(2), pp. 421–30.

Napolitano, Marcos and Wasserman, Maria Clara (2000) "Desde que o samba é samba: a questão das origens no debate historiográfico sobre a música popular brasileira." *Revista Brasileira de Historia*, 20(39), pp. 167–89.

Nascimento, Abdias do (1978) "African Culture in Brazilian Art." *Journal of Black Studies*, 8(4), pp. 389–422.

Nascimento, Abdias do and Nascimento, Elisa Larkin (1979) "Reflections of an Afro-Brazilian." *Journal of Negro History*, 64(3), pp. 274–82. Available at: www.jstor.org/stable/2717038 (date accessed May 11, 2015).

Natali, Marcos (2007) "The Realm of the Possible: Remembering Brazilian Futebol." *Soccer & Society*, 8(2), pp. 267–82.

Oliveira, Francisco de (1985) "Além da transição, aquém da imaginação." *Novos Estudos*, 12, pp. 49–63.

Oliven, Ruben George (1984) "A Malandragem na Musica Popular Brasileira." *Latin American Music Review/Revista de Música Latinoamericana*, 5(1), pp. 66–96.

Ovalle, Priscilla (2008) "Urban *Sensualidad*: Jennifer Lopez, *Flashdance* and the MTV Hip-Hop Re-generation." *Women & Performance: A Journal of Feminist Theory*, 18(3), pp. 253–68.

Pardue, Derek (2002) "Jogada Linguística: Discursive Play and the Hegemonic Force of Soccer in Brazil." *Journal of Sport and Social Issues*, 26(4), pp. 360–80.

Pemberton, John (1975) "Eshu-Elegba: The Yoruba Trickster God." *African Arts*, 9(1), pp. 20–7, 66–70, 90–2.

Perrone, Charles A. and Dunn, Christopher (2002) "Brazilian Popular Music & Globalization." *Journal of Popular Music Studies*, 14(2), pp. 163–5.

Pesavento, Sandra Jatahy (2007) "Uma cidade sensível sob o olhar do 'outro': Jean- Baptiste Debret e o Rio de Janeiro (1816–1831)." *Revista de Historia de Estudos Sociais* 4, 4.

Piedra, José (1991) "Poetics of the Hip." *New Literary History*, 22(3), pp. 633–75.

Pinto, Tiago de Oliveira (1994) "The Pernambuco Carnival and its Formal Organizations: Music as Expression of Hierarchies and Power in Brazil." *Yearbook for Traditional Music*, 26, pp. 20–38.

Prandi, Reginaldo (2004) "Afro-Brazilian Identity and Memory." *Diogenes*, 51, pp. 35–43.

Quijano, A. and Ennis, M. (2000) "Coloniality of Power, Eurocentrism, and Latin America." *Nepantla: Views from South*, 1(3), pp. 533–80.

Quijano, A. and Wallerstein, I. (1992) "Americanity as a Concept. Or the Americas in the Modern World-System." *International Journal of Social Sciences*, 134, pp. 549–57.

Reed, Susan (1998) "The Politics and Poetics of Dance." *Annual Review of Anthropology*, 27, pp. 503–32.

Reis, Daniela (2005a) "Ballet Stagium e o debate sobre a dança moderna brasileira no contexto sócio-político da década de 1970." *Fenix: Revista de História e Estudos Culturais*, 2(1), pp. 1–14.

——. (2005b) "O Balé do Rio de Janeiro e de São Paulo entre as décadas de 1930 e 1940: concepções de identidade nacional no corpo que dança." *Fênix: Revista de História e Estudos Culturais*, 2(3), pp. 1–15.

Reis, João José (1989) "Quilombos e revoltas escravas no Brasil." *Revista USP* 28, pp. 14–39.

——. (1996) "Identidade e Diversidade Étnicas nas Irmandades Negras no Tempo da Escravidão." *Tempo*, 2(3), pp. 7–33.

——. (1997) "'The Revolution of the Ganhadores': Urban Labour, Ethnicity and the African Strike of 1857 in Bahia, Brazil." *Journal of Latin American Studies*, 29, pp. 355–93.

——. (2005) "Batuque: African Drumming and Dance between Repression and Concession, Bahia, 1808–1855." *Bulletin of Latin American Research*, 24(2), pp. 201–14.

Roach, Joseph R. (1992) "Slave Spectacles and Tragic Octoroons: A Cultural Genealogy of Antebellum Performance." *Theatre Survey*, 33, pp. 167–87.

Rosas, R. (2008) "The Gambiarra: Considerations on a Recombinatory Technology." *Caderno Videobrasil*, 2, pp. 36–53. Available at: www2.sescsp.org.br/sesc/videobrasil/up/arquivos/200611/20061117_160212_CadernoVB02_p.36-53_P.pdf (date accessed May 11, 2015).

Rosenn, Keith S. (1971) "The Jeito: Brazil's Institutional Bypass of the Formal Legal System and its Developmental Implications." *American Journal of Comparative Law*, 19(3), pp. 514–49.

Sansone, Livio (1991) "A Produção de Uma Cultura Negra." *Estudos Afro-Asiáticos*, 20, pp. 121–34.

Santos, Boaventura de Sousa (2002) "Between Prospero and Caliban: Colonialism, Postcolonialism, and Inter-identity." *Luso-Brazilian Review* 39(2), pp. 9–43.

——. (2007) "Beyond Abyssal Thinking: From Global Lines to Ecologies of Knowledges." Eurozine, June, pp. 1–33. Available at: http://www.eurozine.com/articles/2007-06-29-santos-en.html (date accessed May 22, 2015).

Santos, Jocélio Teles dos (1994) "As políticas públicas e a população afro-brasileira." *Revista Análise e Dados*, pp. 67–70.

——. (1999) "Nação Mestiça : Discursos e práticas oficiais sobre os afro-brasileiros." *Luso-Brazilian Review*, 36(1), pp. 19–32.

Seigel, Micol and Gomes, Tiago de Melo (2002) "Sabina das Laranjas: gênero, raça e nação na trajetória de um símbolo popular, 1889–1930." *Revista Brasileira de Historia*, 22(43), pp. 171–93.

Shaw, Lisa (2002) "Samba and Brasilidade. Notions of National Identity in the Lyrics of Noel Rosa (1910–1937)." *Lusotopie*, 2, pp. 81–96.

Silva, Renato Rodrigues da (2005) "Hélio Oiticica's Parangolé or the Art of Transgression." *Third Text*, 19(3), pp. 213–31.

Silva, Soraia Maria (2006) "A linguagem do corpo." *Humanidades*, 52, pp. 53–63.

Soares, Carlos, Líbano, Eugênio, and Gomes, Flávio (2001) "'Com o Pé sobre um Vulcão': Africanos Minas, Identidades e a Repressão Antiafricana no Rio de Janeiro (1830–1840)." *Estudos Afro-Asiáticos*, 23(2), pp. 1–44.

Talmon-Cvaicer, Maya (2004) "Verbal and Non-verbal Memory in Capoeira." *Sport in Society*, 7(1), pp. 49–68.

Taylor, J.M. (1982) "The Politics of Aesthetic Debate: The Case of Brazilian Carnival." *Ethnology*, 21(4), pp. 301–11. Available at: www.jstor.org/stable/3773761 (date accessed May 11, 2015).

Thompson, Robert Farris (1966) "An Aesthetic of the Cool: West African Dance." *African Forum*, 2(2), pp. 85–122.

——. (1973) "An Aesthetic of the Cool." *African Arts*, 7(1), pp. 40–3, 64–7, 89–91.

Thornton, John K. (1991) "Legitimacy and Political Power: Queen Njinga, 1624–1663." *Journal of African History*, 32(1), pp. 25–40. Available at: www.jstor.org/stable/182577 (date accessed May 11, 2015).

Treece, David (1997) "Guns and Roses: Bossa Nova and Brazil's Music of Popular Protest, 1958–68." *Popular Music*, 16(1), pp. 1–29.

Vassalo, Simone Ponte (2003) "Capoeiras e intelectuais: a construção coletiva da capoeira 'autêntica'." *Estudos Históricos*, 32, 106–24.

Vassberg, David E. (1976) "African Influences on the Music of Brazil." *Luso-Brazilian Review*, 13(1), pp. 35–54.

Velloso, Mônica Pimenta (1990) "As tias baianas tomam conta do pedaço: espaço e identidade cultural no Rio de Janeiro." *Estudos Históricos*, 3(6), pp. 207–28.

Viveros Vigoya, M. (2002) "Dionysian Blacks: Sexuality, Body, and Racial Order in Colombia." *Latin American Perspectives*, 29(2), pp. 60–77.

Waterman, Christopher (1990) "Our Tradition is a Very Modern Tradition: Popular Music and the Construction of Pan Yoruba Identity." *Ethnomusicology*, 34(3), pp. 367–79.

White, Hayden (1980) "The Value of Narrativity in the Representation of Reality." *Critical Inquiry*, 7(1), pp. 5–27.

Williams, Judith Michelle (2006) "Uma Mulata, Sim! Araci Cortes, 'the Mulatta' of the Teatro de Revista." *Women & Performance: A Journal of Feminist Theory*, 16(1), pp. 7–26.

Wisnik, José Miguel (2000) "O interesse pela dança foi despertado em mim por eles." *Estudos Avançados*, 14(40), pp. 313–20.

——. (2006) "The Riddle of Brazilian Soccer: Emancipatory Dimensions of Culture." *Literature and Arts of the Americas*, 39(2), pp. 198–209.

——. (2011) "Ambivalence, Paradox and the Poison-Remedy of Brazilian Improvisation." *Critical Studies in Improvisation*, 7(1). Available at: www.criticalimprov.com/rt/printerFriendly/1367/2027#_edn15 (date accessed May 11, 2015).

Yan, Haiping (2005) "Other Transnational: An Introductory Essay." *Modern Drama: World Drama from 1850 to the Present*, 48(2), pp. 225–48. Available at: http://muse.jhu.edu/journals/modern_drama/summary/v048/48.2yan.html (date accessed May 11, 2015).

Magazine articles

"A Capoeira" (1906) *Revista Kosmos III*, Rio de Janeiro, March.

Anauate, Gisela (2007) "Beleza Brutal." *Revista Época*, August.

Andreata, Valeria (2007) "Nas coreografias do Grupo Corpo, ex-bailarino dá arte a vida." *Revista do Minas*, June.

Bragato, Marcos (n.d.) "Rodrigo Pederneiras." *Dançar Express*.

Cabral, Sérgio et al. (1971) "Madame Satã." *O Pasquim*, April 29–May 5, p. 2.

Katz, Helena. (n.d.) "Sucessos, de Minas para o mundo." *Noticias Shell*.

Machado, Elmar. (1976) "Madame Satã para O Pasquim: 'Enquanto eu viver, a Lapa viverá'." *Pasquim* 357, April 30.

Morgan, Clyde Wesley. (1976) "Danças e ritmos negors no Brasil." *Cultura*, December.

Pastinha, Vicente Ferreira (2006) "Capoeira: Brazilians Karate." *Black Belt Magazine*, July 20. Available at: www.portalcapoeira.com/Publicacoes-e-Artigos/documento-historico-capoeira-brazilians-karate (date accessed May 11, 2015).

Pinto, Aina (2007) "A maratona do Corpo." *ISTOÉ*, August.

Ponzio, Ana Francisca (1994) "Let's Face the Music and Dance." *Dance Magazine*, March.

Schild, Susana (n.d.) "Enigma desvendado." *Noticias Shell.*

Thompson, Robert Farris (1988) "Tough Guys Do Dance." *Rolling Stone*, March 24.

Viana, Edilene (2007) "Novo Espetáculo do Grupo Corpo estrea na França." *Total Magazine*, October.

Vieira, Luis Renato (1999) "Waldemar da Paixão, o Mestre da Liberdade." Capoeira Magazine.

Newspaper articles

Ambrosio, Marcelo (1991) "Corpo a corpo." *Jornal de Brasília*, November 14.

Andreas, Adriano (1993) "Grupo Corpo Homenagea Ernesto Nazareth." *A Folha da Imprensa*, September 17.

Araujo, Celso (1992) "Grupo Corpo continua na crista da onda." *Jornal de Brasília*, July 11.

Avellar, Larcello Castilho (2007) "O Corpo e a Crise." *Estado de Minas*, August 4.

Avellar, Marcello Castilho (1992) "Corpo descobre o tempo nas pulsações do espaço." *Estado de Minas*, June 26.

"Brazilian dancers make bay debut." (1994). *Herald Express*, October 7.

Cavalcanti, Nireu (1999) "O capoeira." *Jornal do Brasil*, November 15. Available at: www.capoeira-palmares.fr/histor/nireu.htm (date accessed May 11, 2015).

Comodo, Roberto (1993) "Meneios Cariocas." *Jornal do Brazil*, April 26.

Cozer, Raquel (2007) "Corpo Bruto." *Folha de São Paulo*, July 23.

Crisp, Clement (1994) "Smiles, Sambas, and Psychotic Behavior." *Financial Times*, October 15.

Ferreira, Adriana (1992) "Sortilegio mineiro no corpo." *O Globo*, October 8.

Freyre, Gilberto (1938) "Foot-ball mulato." *Diario de Pernambuco*, June 17, p. 4.

Fry, Peter (1995) "Why Brazil is Different." *Times Literary Supplement* (December).

Guzik, Alberto (1993) "O jeito Mineiro de Nazareth." *Jornal da Tarde*, August 24.

Heliodora, Barbara (n.d.) "Grupo Corpo coreografa uma nova forma de beleza." *O Globo*.

Joory, Eva (1993) "O prazer da Brasilidade." *O Globo*, 3 May.

Jowitt, Deborah (2008) "Grupo Corpo and Emanuel Gat: Doing Reps." *The Village Voice*, 1 April, p. 1. Available at: www.villagevoice.com/2008-04-01/dance/doing-reps (date accessed May 11, 2015).

Katz, Helena (2007a) "Seco e violento, Corpo brilha no breu." *O Estado de São Paulo*, August 3.

——. (2007b) "Dentro do Mais Puro *Breu*." *O Estado de São Paulo*, July 23.

Keavy, Hubbard (1942) "Eros Volusia, Brazilian Dancer, Makes Her Screen Debut Soon." *Pittsburgh Post-Gazette*, January 3. Available at: http://news.google.com/newspapers?nid=1129&dat=19420103&id=tdkOAAAAIBAJ&sjid=CGoDAAAAIBAJ&pg=4879,6639891 (date accessed May 11, 2015).

Kisselgoff, Anna (1994) "Lyons Biennial Focuses on African Influences." *New York Times*, September 28, Dance Review.

Linhote, Leandro (2007) "*Breu*: Musica da Escuridão." *O Globo*, July 27.

Magalhaes, Geraldo (1993) "Gestos que ofuscam." *Estado de Minas*, July 4.

Mascagna, Silvana (2007) "O homem Rasteja; o Corpo, não." *Magazine*, September 12.

Meira, Tatiana (2008) "Capoeira consagra sua ginga." *Diario de Pernambuco*, July 16.

Navas, Cassia (1993) "Corpo promove dialogo entre o erudito e o popular." *Folha de São Paulo*, May 7.

Pacini, Paulo (2009) "A Rua da Imperatriz." *Jornal do Brasil*, October 30. Available at: www.jblog.com.br/rioantigo.php?itemid=16928 (date accessed May 11, 2015).

Paladino, Patrícia (n.d.) "Corpo em evidencia." *O Globo*.

Palomino, Erika. (n.d.) "Corpo faz revoluções em sua trajetória." *O Globo*.

——. (1992a) "Corpo Brinca com números em novo balé." *Folha de São Paulo*, June 9.

——. (1992b) "Corpo dança o numero 21 em estreia hoje em SP." *Folha de São Paulo*, June 18.

——. (1993) "Grupo Corpo volta com rebolado brasileiro." *Folha de São Paulo*, April 21.

Paulo, João (2007) "Luz sobre o breu cotidiano." *Estado de Minas*, September 15.

Pelegrini, Liliane (2007) "Oembate num bale impactante." *Pampulha*, September 7.

Pereira, Roberto (2007) "Contruindo um novo vocabulário." *Jornal do Brasil*, August 17.

Pimenta, Angela (1993) "Entre o Pudor e o Pecado." *Veja*, April 28.

Pinheiro, A. (2008) "Mestiçagem latino-americana." *Jornal o Povo*. May 10. Available at: http://barroco-mestico.blogspot.com/2008/05/entrevista-do-amlio-para-o-jornal-o.html (date accessed May 11, 2015).

Piza, Daniel (2007) "O coreógrafo do ritmo brasileiro." *O Estado de São Paulo*, August 8.

Ponzio, Ana Francisca (1992) "Poder ilimitado de renovação." *O Estado de São Paulo*, June 20.

——. (1993a) "'Corpo' faz literatura entrar na dança." *O Estado de São Paulo*, April 22.

——. (1993b) "Corpo se renova com Nazareth." *O Estado de São Paulo*, April 22.

——. (1994) "Corpo Conquista critica e publico em Lyon." *Folha de São Paulo*, September 26.

Reis, Sérgio Rodrigo (2007) "Caos e cosmos." *Estado de Minas*, July 21.

Riding, Alan (1996) "A Dance Festival Marches to a Different Drummer." *New York Times*, September 18. Available at: www.nytimes.com/1996/09/18/arts/a-dance-festival-marches-to-a-different-drummer.html (date accessed May 11, 2015).

Rocha, Pedro (2008) "Mestiçagem latino-americana: entrevista com Amalio Pinheiro." *O Povo*, May 10.

Rockwell, John. (2005) "Fast and Sometimes Tricky Moves, Straight Out of Brazil." *New York Times*, October 27. Available at: www.nytimes.com/2005/10/27/arts/dance/27grup.html?scp=1&sq=grupo%20corpo%20rockwell&st=cse (date accessed May 22, 2015).

Sampaio, Flavia (2007) "A música como fio locomotor." *Jornal do Estado*, November 22.

Schwarcz, Lilia Moritz (2006) "A Mestizo and Tropical Country: The Creation of the Official Image of Independent Brazil." *Revista Europea de Estudios Latinoamericanos y del Caribe*, April 30, pp. 25-42.

Sebastião, Walter (1993) "O Grupo Corpo dança a ginga brasileira." *Estado de Minas*, June 2.

Soter, Silvia (2007) "Estranhamento e fricção num caldeirão de referências urbanas." *O Globo*, August 18.
Sulcas, Roslyn (2008) "Exotica of Brazil in Motion." *New York Times*, March 27. Available at: www.nytimes.com/2008/03/27/arts/dance/27grup.html?_r=1andref=dance (date accessed May 11, 2015).
"Um coreografo em busca do desconhecido." (n.d.). *O Globo*.
Veloso, Caetano (1991) "Carmen Miranda Dada." *Folha de São Paulo*, December 13. Available at: http://tropicalia.com.br/leituras-complementares/carmen-miranda-dada#sthash.GzPGkywD.dpuf (date accessed May 11, 2015).
Wisnik, José Miguel (1993) "Nazareth é espelho de Machado de Assis." *Folha de São Paulo*, April 21.

Papers presented at a meeting or conference

Acordeon, Mestre (n.d.) "Brief Description of Capoeira." Unpublished paper.
Alencastro, Luiz Felipe de (2006) "Articulação das Idéias Raciais no Mundo Ibérico." Public lecture given at the *Fábrica de Idéias* Conference. Salvador: CEAO.
Carvalho, José Jorge de Carvalho (1996) "Predicaments of Black Maculinity in Brazil as Seen through Popular Culture." Working Paper presented to the *Rockefeller Program on Afro-American Identity and Cultural Diversity*, Center for Latin American Studies, University of Brasilia-UnB, Brazil.
Dantas, Danilo Fraga (2005) "A dança invisível: sugestões para tratar da performance nos meios auditivos." Unpublished.
De Sá, Bibiana Gutierrez Fernandes (2008) "Grupo Corpo – Cultura Nacional na Dança Contemporânea." *IXXXI Congresso Brasileiro de Ciências da Comunicação*. Natal, September 2–6.
Oliveira. A.L. de (2002) "Os significados dos gestos n jogo da capoeira: aspectos históricos e sociais." *Anais: Encontro de História, do Esporte, Lazer e Educação Física*.
Pantoja, Selma (2000) "Quitandas e quitandeiras: história e deslocamento na nova lógica do espaço em Luanda." *Atas da III Reunião Internacional de História de África*. Lisbon.
Sansone, Livio (1999) "From Africa to Afro: Use and Abuse of Africa in Brazil."Amsterdam/Dakar: SEPHIS, CODESRIA.
Santos, Boaventura Sousa (2003) "Sociologia das ausências e a sociologia das emergências." *XI Congresso Brasileiro de Sociologia na UNICAMP*. Campinas-SP, Brasil, September 3.
Santos, Juana Elbein dos (2003) "Bimaster Didi: Tradition and Contemporaneity." *XXIII Bienal Internacional de São Paulo*.
Schwarcz, Lilia (1989) "De festa também se vive. Reflexoes sobre o centenário da abolição em São Paulo." Unpublished paper. Rio de Janeiro.
Tavares, Julio Cesar (2002) "Embodied Language, Gesture and Memory: How Can One 'Get Ginga'." *International Society for Gesture Studies Proceedings*. Austin, TX, June 5–8. Available at: www.utexas.edu/coc/cms/International_House_of_Gestures/Conferences/Proceedings/Contributions/deTavares/deTavares.html (date accessed May 21, 2015).

Theses and dissertations

Bezerra, Riselia Duarte (2000) "Sambations: Samba and the Politics of Syncopation." PhD dissertation, University of California, Riverside.

Bragato, Marcos (1996) "Saltos, emergências, permanências: três tempos na obra de Rodrigo Pederneiras." Master's thesis, PUC/USP.

Carvalho, Deliane da Silva (1993) "Eleições no Segundo Reinado os capoeiras no Rio de Janeiro." Master's thesis, IFCH, UERJ.

Conrado, Amélia Vitória de Souza (1996) "Dança étnica afro-baiana: uma educação movimento." Master's thesis, Universidade Federal da Bahia.

Desch-Obi, Thomas J. (1994) "Knocking and Kicking, Ladya, and Capoeira: Resistance and Religion in the African Martial Arts of the Black Atlantic." Master's thesis, University of California, Los Angeles.

——. (2000) "Engolo: Combat Traditions in African and African Diaspora." PhD dissertation, University of California, Los Angeles.

Dossar, Kenneth (1994) "Dancing between Two Worlds: An Aesthetic Analysis of Capoeira Angola." PhD dissertation, Temple University.

Downey, Greg (1998) "Incorporating Capoeira: Phenomenology of a Movement Discipline." PhD dissertation, University of Chicago.

Monteiro, Marianna F. Martins (2002) "Espetáculo e devoção: burlesco e teologia-política nas danças populares brasileiras." PhD dissertation, Universidade de São Paulo.

Paixão, Paulo (2009) "Por Uma politica cidada do corpo: A funcão communicative do nacionalismo na danca no Brasil." PhD dissertation, PUC, São Paulo.

Reis, Letícia V. de Souza (1993) "Negros e brancos no jogo da capoeira: a reivenção da tradição." Master's thesis, FFLCH-USP.

Tavares, Júlio César de Souza (1984) "Dança da guerra: arquivo-arma." Master's thesis, Universidade de Brasilia.

——. (1998) "Gingando and Cooling Out: The Embodied Philosophies of the African Diaspora." PhD dissertation, University of Texas.

Vieira, Luiz Renato (1991) "Vadiacão a capoeira regional: uma interpretacão da modernizacão cultural no Brasil." PhD dissertation, UnB.

Playbills and pamphets

Araújo, Rosangela (1993) "GCAP: 10 Anos Gingando na Mesma Luta." Pamphlet.

Bimba, Mestre (1960s) "Curso de capoeira Regional." [LP leaflet] Salvador-BA.

Burlamaqui, Anibal (1928) "Gymnástica Nacional (Capoeiragem) Methodizada e Regrada." Pamphlet.

Grupo Corpo: Brazilian Dance Theatre (2009) "Seven or Eight Pieces for a Ballet and *Breu*." Playbill, Jackson Hall, Mondavi Center, UC Davis.

Katz, Helena and Wisnik, José (1993) "Nazareth." Playbill, Grupo Corpo.

Nova Lei Rouanet: Ampliar e qualificar o Acesso aos Recursos, Financiar Todas as Dimenções da Cultura, em Todas as Regiões do Brasil (2010) Brasilia: Ministry of Culture. Available at: http://blogs.cultura.gov.br/blogdarouanet/category/publicacoes (date accessed November 2, 2012).

Vacchiano, Anna (ed.) (2007) "Vedetes em revista: teatro de revista: exposição de 09 de fevereiro de 2007 a 06 de maio de 2007." Pamphlet. São Paulo, Caixa Cultural.

Van-Dúnem, Domingos (1987) *Sobre o vocábulo kitandeira*. Pamphlet.

Websites

Antonil, André João (1999) "Como se há de haver o senhor do engenho com seus escravos." Jangada Brasil, May. Available at: www.jangadabrasil.com.br/maio/fe90500c.htm (date accessed May 11, 2015).

Araujo, Clelia (2008)"Capoeira é registrada como patrimônio imaterial brasileiro." Ministério da Cultura, July 16. Available at: www.cultura.gov.br/site/2008/07/16/capoeira-e-registrada-como-patrimonio-imaterial-brasileiro (date accessed May 11, 2015).

ASCOM/IPHAN (2007) "Samba do Rio de Janeiro é Patrimônio Cultural do Brasil." Ministerio da Cultura, October 10. Available at: www.cultura.gov.br/site/2007/10/10/samba-do-rio-de-janeiro-e-patrimonio-cultural-do-brasil (date accessed May 11, 2015).

Association de Capoeira Palmares de Paris (2007) "Capoeira: Historical Sources Timeline." September 11. Available at: www.capoeira-palmares.fr/histor/index_en.htm (date accessed May 11, 2015).

Assunção, Matthias Röhrig and Mestre Cobra Mansa (2009) "A dança da zebra." Portal Capoeira, October 2. Available at: www.portalcapoeira.com/Publicacoes-e-Artigos/a-danca-da-zebra (date accessed May 11, 2015).

Briand, Pol and Lucia Palmares (n.d.) "Machado de Assis." Association de capoeira Palmares de Paris. Available at: www.capoeira-palmares.fr/histor/machado.htm (date accessed May 11, 2015).

"Capoeira Ginga Nâgo." (n.d.) Available at: www.ginganago.com (date accessed May 11, 2015).

"Código Penal: Decreto nº 847 de 11 de Outubro de 1890." (n.d.) Centro internacional de estudos e pesquisas sobre a infância. Available at: www.ciespi.org.br/base_legis/legislacao/DEC20a.html (date accessed May 11, 2015).

Debret, Jean-Baptiste. (n.d.) "Valongo Mercado de Vidas." Jangada Brasil. Available at: www.jangadabrasil.com.br/maio/of90500b.htm (date accessed May 11, 2015).

Evelin, M., Lima, D., et al. (n.d.) "Coleção Brasilidade apresenta: o gesto." Available at: www.nucleododirceu.com.br/colecao-brasilidade-apresenta-o-gesto (date accessed September 4, 2013).

Handler, Jerome S. and Tuite, Jr., Michael L. (2010) "The Atlantic Slave Trade and Slave Life in the Americas: A Visual Record." *Virginia Foundation for the Humanities*, April 5. Available at: http://hitchcock.itc.virginia.edu/Slavery/index.php (date accessed May 11, 2015).

"Jongos, Calangos e Folias: memória da música negra no Rio de Janeiro-Pesquisa Bibliográfica." (n.d.) Universidade Federal Fluminense. Available at: www.historia.uff.br/jongos/?page_id=67 (date accessed May 11, 2015).

Lobo, Carol. "Capoeira: Patrimônio Brasileiro. Manifestação cultural está inscrita nos Livros dos Saberes e das Formas de Expressão." Ministério da Cultura. O dia-a-dia da Cultura. Available at: www.cultura.gov.br/site/2008/07/16/capoeira-patrimonio-cultural-brasileiro (date accessed May 11, 2015).

"Michaelis Moderno Dicionário da Lingua Portuguesa." (n.d.) Available at: http://michaelis.uol.com.br/moderno/portugues/index.php (date accessed May 11, 2015).

Pantoja, Selma (n.d.) "Conexões e identidades de gênero no caso Brasil e Angola, Séculos XVIII–XIX." Casa das Africas, Available at: http://webcache.googleusercontent.com/search?q=cache:5gl9UQ5W8r0J:www.casa

dasafricas.org.br/site/img/upload/415151.rtf+Conex%C3%B5es+e+identidades+de+g%C3%AAnero+no+caso+Brasil+e+Angola+,+S%C3%A9cs.+XVIII-XIX&cd=1&hl=en&ct=clnk&gl=us&client=firefox-a (date accessed May 11, 2015).
"Rio de Janeiro: uma nova ordem na cidade: Limpeza do pântano do Valongo." (1815) O Arquivo Nacional e a História Luso-Brasileira. Available at: www.historiacolonial.arquivonacional.gov.br/cgi/cgilua.exe/sys/start.htm?infoid=849andsid=103andtpl=printerview (date accessed May 11, 2015).
Schwartz, Wagner (2014) "Transobjeto." Available at: www.wagnerschwartz.com/transremontagem.html (date accessed May 11, 2015).
——— (2006) "Pré-texto." Available at: www.wagnerschwartz.com/transremontagem.html (date accessed May 11, 2015).
Sucato, Steve (2008) "Premium *Breu* – Grupo Corpo Unveils the US Premiere of its New Work." Explore Dance, March 22. Available at: www.exploredance.com/article.htm?id=2218 (date accessed May 11, 2015).
Wisnik, José Miguel (n.d.) "Music – Brazilian Embassy." Embassy of Brazil. Available at: www.brasilemb.org/culture/music (date accessed May 11, 2015).

Recordings: LPs, videos and DVDs

A relacão entre musica e danca no grupo corpo: Round table with Gilberto Assis, Ines Borgea, Ze Miguel e Ze Luis Martins (2002) [DVD].
Banana da Terra (1939) Ruy Costa [Film]. Rio de Janeiro and São Paulo.
Bahia: Africa in the Américas (1988) Geovanni Brewer. [DVD]. UCCMIL.
Bimba, Mestre (1960s) *Curso de capoeira regional.* [LP] Salvador: JS discos.
Black Orpheus (2006) Marcel Camus [DVD]. USA: Image Entertainment.
Capoeira Angola do Mundo (2003) Capoeira Angola Center of Mestre João Grande [DVD]. New York.
Capoeiragem' in Bahia.(2000) José Umberto [VHS]. Bahia: IRDEB/TVE.
Capoeirando ilheus (2004) Mestre Suassuna [DVD]. Manaus: Sonopress.
Carmen Miranda: Bananas is My Business (1995) David Meyer and Helena Solberg [DVD]. Fox Lorber Home Video.
Carnaval 2008 (2008) Aloysio Legey [DVD]. Globo Marca e Som Livre.
Deusas Do Ébano: 24 Noite da Beleza Negra – Ile Aiye (n.d.) Angela Figueiredo [DVD]. Salvador.
The Gang's All Here (1943) Busby Berkeley [Film]. USA: Twentieth Century Fox Film Corporation.
Ginga: The Soul of Brazilian Football (2005) Tocha Alves, Hank Levine and Marcelo Machado [Film]. O2 Filmes, Nike, Wieden and Kennedy.
——. (2014) Tocha Alves, Hank Levine and Marcelo Machado. [DVD]. Mr Bongo Films.
Grupo Corpo 30 anos: uma família brasileira (2007) Fábio Barreto e Marcelo Santiago. [DVD]. Brasil: Paramount.
Grupo Corpo Presents: 21 (2005) Paulo Pederneiras. [DVD].
Grupo Corpo Presents: Nazareth (2005) Paulo Pederneiras [DVD].
Grupo Corpo Presents: Sete ou oito peças para um ballet (1994). Paulo Pederneiras and Fernando Velloso. [DVD].
Grupo Corpo: Companhia de Dança (2004) Jürgen Wilcke and produced by Hessischer Rundfunk [DVD]. West Long Branch: Kultur International Films.

Jongos, Calangos, e Folias: Musica Negra, memoria e poesia (2008) Hebe Mattos and Martha Abreu. [DVD]. Rio de Janeiro: LABHOI e NUPEHC.

Moro no Brasil (2006) Mika Kaurismäki; Mika Kaurismäki and George Moura. [DVD]. Burbank: Milan Records.

Musica dos Mestres (ndn) Almeida, Ubirajara, and Kelly Calderon, vol. 1 [DVD]. USA.

O Pulo do Gato (2002) Capoeira Angola Center of Mestre João Grande. [DVD]. New York.

O Samba: Warriors of Dance (2005) Jean Claude Guiter. [DVD]. Miami: Synkronized.

O'Gun Ye: Kamala Ye (2007) Joice Rodrigues. [DVD]. Salvador: Fundação Internacional Capoeira Angola.

Only the Strong (1993) Sheldon Lettich. [DVD]. Miami: Davis-Film.

Pastinha! Uma Vida pela Capoeira (1999) Antonio Carlos Muricy. [VHS].

Piracema (2011) Lia Rodrigues. France: Mercat de les Flors. Available at: https://www.youtube.com/watch?v=GQ77JOXOA64 (date accessed May 11, 2015).

Pororoca (2010) Lia Rodrigues. Festival Panorama (November 5–21). Available at: www.youtube.com/watch?v=eMmFvCXdx3g (date accessed May 11, 2015).

The Spirit of Samba: The Black Music of Brazil (2001) Jeremy Marre. [DVD]. Shanachie Entertainment Corp.

That Night in Rio (1941) Irving Cummings. [Film]. USA: Twentieth Century Fox Film.

Tropicália 2 (1994) Caetano Veloso e Gilberto Gil [CD]. Elektra/Nonesuch.

Vem jogar mais eu, meu irmão (n.d.) Grupo de Capoeira Angola Ypiranga de Pastinha. [DVD]. Paris: Helico Presente.

Index

9781137462268